GAS PHASE CHROMATOGRAPHY

ENGLAND:	BUTTERWORTH & CO. (PUBLISHERS) LTD. LONDON : 88 Kingsway, W.C.2
AFRICA:	BUTTERWORTH & CO. (AFRICA) LTD. DURBAN : 33/35 Beach Grove
AUSTRALIA :	BUTTERWORTH & CO. (AUSTRALIA) LTD. SYDNEY : 6/8 O'Connell Street MELBOURNE : 473 Bourke Street BRISBANE : 240 Queen Street
CANADA :	BUTTERWORTH & CO. (CANADA) LTD. TORONTO : 1367 Danforth Avenue, 6
NEW ZEALAND :	BUTTERWORTH & CO. (NEW ZEALAND) LTD. WELLINGTON : 49/51 Ballance Street AUCKLAND : 35 High Street
U.S.A.:	BUTTERWORTH INC. WASHINGTON, D.C. : 7235 Wisconsin Avenue, 14

GAS PHASE CHROMATOGRAPHY

Volume I
GAS CHROMATOGRAPHY

RUDOLF KAISER

Badische Anilin- und Sodafabrik AG Ludwigshafen

Translated by
P. H. Scott

LONDON
BUTTERWORTHS
1963

ISBN 978-1-4684-8293-5   ISBN 978-1-4684-8291-1 (eBook)
DOI 10.1007/978-1-4684-8291-1

Originally published under the title

Chromatographie in der Gasphase
Band I Kapillar-chromatographie

by Bibliographisches Institut AG Mannheim

© 1960 Bibliographisches Institut AG Mannheim
Softcover reprint of the hardcover 1st edition 1960

FOREWORD

THE present volume, which is the first of a three-volume work on gas phase chromatography, deals with the problems of gas chromatography in packed columns.

Gas chromatography, like any other analytical method, is mainly a matter of practical skill, and therefore emphasis has been given to the apparatus at the expense of a more detailed presentation of the theory. The aim of this book is to make lecturers and students, chemists, works engineers and laboratory workers familiar with this highly effective branch of analytical physical chemistry. I hope too that the experienced worker may find references which will be of value to him in his work and which will spare him part of the now almost impossible task of keeping up to date with the literature.

The nomenclature used here is the result of a number of discussions with Professor E. Cremer and Dr. E. Bayer, and I should like to take this opportunity of expressing my grateful thanks to them.

The present book is based partly on my book *Gas Chromatography* which appeared at the end of 1959. Numerous discussions with Professor E. Leibnitz and his colleagues H. P. Angele, M. Hofmann, H. Holzhäuser, M. Kuhl and H. G. Struppe and the experimental work carried out with them have all influenced this revision.

I should also like to thank Dr. H. Kienitz and his colleagues Dr. K. Dorfner, Dr. H. D. Ermshaus and Dr. H. Runge for valuable suggestions.

Beckmann Instruments, GmbH, Munich, Perkin-Elmer & Co., Bodenseewerk, Überlingen, Griffin & George Ltd., Alperton, Wembley, W. G. Pye & Co. Ltd., Cambridge, and the organizers of the 3rd Gas Chromatography Symposium at Edinburgh in 1960 have all been most generous in supplying me with data and illustrative material.

I should also like to thank the Verlag Bibliographisches Institut, Mannheim, for their splendid co-operation.

RUDOLF KAISER

CONTENTS

Contents

INTRODUCTION

DURING recent years gas chromatography has developed into the most effective of the chromatographic analytical methods. With its aid qualitative and quantitative separations and determinations can be carried out on mixtures of any substances which vaporize without decomposition under the conditions of gas chromatography or which can be decomposed in a reproducible manner. This includes all organic and inorganic substances with definite boiling points, all substances capable of being converted into stable end-products, and all those substances which, although originally not covered by these terms, can be converted into vaporizable or decomposable derivatives.

Gas chromatographic methods can be used for automatic operation. This is why their use in works laboratories, pilot plants, and full-scale chemical plants is widespread and still increasing. Modern gas chromatographic methods require only a thousandth to a millionth of a gram of substance for an analysis, and can be completed within minutes or even in seconds. The new method provides a valuable extension of the range of analytical techniques, not only for the chemist but also for the biologist and the medical man.

The present volume presents the theory of gas chromatography with packed columns, the so-called classical method, in a form which is very simple but is useful for the practician.

A good understanding of the apparatus used in gas chromatography is essential, since in many cases it is necessary to build one's own or to modify existing equipment; in order to assist this and also to help select the most suitable commercial instrument for the job in hand, Volume I gives a very thorough treatment of apparatus problems.

Finally the qualitative and quantitative application of gas chromatography to a few selected practical examples is described in a systematic and generalized manner.

Volume II, which will follow later, will deal with the theory, apparatus and application of capillary gas chromatography. The latter method was only recently developed from classical gas chromatography as a new and in many cases very efficient variant.

Finally the third volume will contain in tabular form all the important data which are necessary or useful for the widest possible application of classical and capillary gas chromatography for qualitative and quantitative analysis, and also for industrial gas chromatography.

Survey of the Four Fundamental Chromatographic Methods

The term 'chromatography' is at present understood to include all those processes in which separation is brought about by adsorption or solution partition of a mixture between two non-miscible phases.

1   1

The two phases flow in intimate contact countercurrent to one another, or one phase is stationary and the other phase flows intensively through it. For the mobile phase only two states of matter can be considered: liquid or gas. Only liquid or solid substances may be used for the stationary phase.

Mobile phase	*Stationary phase*
G gas	L liquid (liquidus)
L liquid	S solid (solidus)

By combinations of these the four possible chromatographic methods (abbrevated to C) are obtained:

$$\left.\begin{array}{l} \text{GLC} \\ \text{GSC} \end{array}\right\} \text{Gas chromatography}$$

$$\left.\begin{array}{l} \text{LLC} \\ \text{SLC} \end{array}\right\} \text{Column, paper, and partition chromatography, and ion exchange.}$$

The following table shows when and by whom the individual methods were first discovered.

SLC	Tswett (1906)
GSC	Schuftan (1931), or Ramsey (1905)
LLC	Martin, Synge (1941)
GLC	James, Martin (1952) or Damköhler, Theile (1943)

There are four techniques by which the separation can be carried out:

1. By elution (elution analysis)
2. By displacement with the mixture itself (frontal analysis)
3. By displacement with an auxiliary substance (displacement analysis)
4. By circulation of a temperature field (circulating chromathermography)

Almost all the chromatographic methods can be carried out by the four techniques:

GLC: Elution, frontal and circulating chromathermography
GSC: Elution, frontal, displacement and circulating chromathermography
LLC: Elution, frontal and displacement analysis
SLC: Elution, frontal and displacement analysis

Modern techniques, however, favour the elution process, because it is much easier to control than frontal and displacement analysis.

Limits of Application, Definitions and Methods of Gas Chromatography

The general term 'gas chromatography' includes all those chromatographic processes in which the essential material transport occurs in the gaseous or vapour phase.

The applicability of gas chromatography is limited to those substances which may be vaporized under normal conditions or which may be decomposed in a reproducible manner to give stable gaseous products. It may also be applied to substances which, although not capable of being vaporized or converted into gaseous products in their original form, can be converted by given chemical reactions into substances with the required properties.

The range of application thus includes all gases, all liquid and solid substances capable of being distilled, all substances which may be decomposed in a reproducible manner such as polycondensates, polymerizates, salts of non-volatile acids, organic complexes of many metals, halides of the transition elements, etc.

We can therefore make the following definition:

Gas chromatography is a rapid process for the separation and analysis of all substances which can be vaporized or decomposed, or which can be chemically converted into such substances. By this means qualitative and quantitative analytical results can be obtained.

There are several gas chromatographic processes; they may be distinguished on the one hand according to the principle of separation, on the other according to the method of operation.

There are two different principles of separation:

1. *Gas-solid chromatography (GSC) or adsorption gas chromatography.* This principle depends on the variation in the extent to which the constituents of a mixture are adsorbed on an adsorbent such as activated carbon, silica gel, alumina, clay, molecular sieves, etc.

2. *Gas-liquid chromatography (GLC) or partition gas chromatography.* This principle depends on the varying solubilities of the vapours of the constituents in the liquid phase, which may be a substance such as tricresyl phosphate, paraffin oil, glycerine, etc.

Each of these principles can be carried out at either constant or fluctuating temperature. The operating temperature may rise steadily or discontinuously, but a rising or falling temperature field can also recur in a cyclic fashion. We therefore speak of *isothermal gas chromatography or gas chromathermography.*

Further, we may distinguish four different methods of operation:

(a) *The elution or development technique.* In this method the mixture is 'transported' through a tube, or separating column, packed with adsorbent or impregnated with liquid phase, by means of a so-called 'carrier gas', which flows continuously and steadily. As a result of this continuous transport of material through the column, the different rates of travel of the individual constituents effect the desired separation.

The essential thing about this method is that the substance need only be introduced into the column inlet on one occasion, at the start of the analysis, and that only a small amount of substance is required.

(b) *The displacement technique.* The continuously flowing carrier gas is saturated with the vapour of a substance which is adsorbed by the column packing more strongly than the constituents of the mixture already adsorbed on it. This leads to the progressive displacement of these substances, which are then removed by the carrier gas. In this method also the sample is introduced only once at the start of the process and only a relatively small quantity is used.

(c) *Frontal analysis.* No carrier gas is needed. In contrast to methods (a) and (b) the test substance is not introduced on one unique occasion, but is passed through the column continuously. The most rapidly travelling components emerge first (as gases) at the foot of the column, then come the slower ones, until finally at the end of the process the mixture is coming out of the column with its original composition.

(d) *Circulating chromathermography.* Whether or not a carrier gas is required depends on the conditions which prevail during the analysis of a given mixture. The substance flows continuously in an unbroken stream through a column which is generally ring-shaped, and which has the column outlet and inlet next to one another. Through the column runs a temperature field in the direction of the flow of substance. This field is short relative to the length of the column but steep. The instant at which the temperature field passes the column outlet on each circuit is equivalent to the end and the beginning of an analysis or an analytical cycle. The detector registers peaks similar to those obtained by the elution technique. The chromatographic picture corresponds to that of a differential chromatogram when, for example, a thermal conductivity cell is used.

Not all the methods can be applied to the two basic processes.

For gas solid chromatography: techniques a, b, c, d.

For gas liquid chromatography: techniques a, d.

The next important factor is the operating temperature and its increase or decrease.

We may distinguish methods which use isothermal, discontinuous, and linearly increasing temperatures.

If we consider all the possible methods according to their importance, we find that only a few of the variants can be used with success:

(a) Gas liquid chromatography using the elution technique, with preferably isothermal temperature increase, but in recent years also with discontinuous temperature programming. It is known as partition gas chromatography, and is abbreviated to GLC;

(b) Gas solid chromatography using the elution technique, with isothermal or linearly increasing temperature programming. It is known as adsorption gas chromatography, and is abbreviated to GSC;

(c) Circulating chromathermography.

Nomenclature

In order to avoid misunderstandings due to the already widespread use of different systems of terms and symbols in gas chromatography, it would be helpful if a standard nomenclature could be devised.

Nomenclature

Apparatus

The

sample injector injects the substance into the
column, which separates the mixture and which is filled with the
column packing. This consists of an
active solid (GSC) or the
solid support which is impregnated with the
liquid phase. Through the column flows the
carrier gas, which transports the separated substances to the
detector. Here electrical signals are produced which are taken up by the
recorder and/or the
integrator. We may distinguish between the
integral detector, which produces a
step chromatogram, and the
differential detector, which produces the ordinary
gas chromatogram or *chromatogram*, which consists of a series of
peaks rising above the
base line.

Evaluation of a Gas Chromatogram

The methods of evaluating chromatograms and the corresponding nomenclature give rise even today to varying interpretations and this is discussed at almost every important technical symposium. With regard to qualitative evaluation, the following points should be kept in mind.

The qualitative analytical results are obtained in the first place as time values. From the time when the sample is injected to the instant of its appearance at the detector the substance to be separated may be regarded as being within the apparatus, i.e. within the column and the connecting parts. This period of retention in the apparatus is composed of two time values: the time that the substance is in the liquid phase or the active solid, and the time when it is passing through the gas piping and the gas in the column.

The time that the substance spends in the gas phase is the same for all substances and thus is not in the least characteristic for any given substance, but depends only upon the conditions imposed by the apparatus. It is thus the time spent in the stationary phase, i.e. in the liquid phase or the active solid, which is characteristic for a given substance, and for this reason it forms the sole basis for analytical evaluation. We must make sure that the names and symbols given to these time values are clearly distinguished, especially in view of the fact that it is the total time value which is always obtained first, the characteristic values only being obtained by calculation.

5

The fact that the difference between these two values is not always very large does not make any difference; simplifications of this kind are at the root of many misunderstandings.

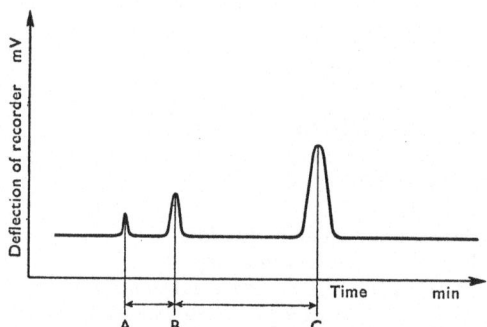

Figure 1. Nomenclature. A = start of analysis. B = air peak. C = substance peak

The following definitions take into account the fact that the total time value and the characteristic time value must be distinguished (see *Figure 1*).

Gas hold-up time t_d (min) corresponds to the distance AB in *Figure 1*
= time from sample injection to appearance of air peak (He, air, H_2;)
= the time spent purely in the gas phase, i.e. the time taken to pass through the gas volume of the column and connecting pieces.

Retention time (uncorrected) t_{dr} (min) corresponds to distance AC in *Figure 1*
= total time for the emergence of the peak maximum after injection of the sample;
= total time spent in the liquid phase or on the active solid plus the time spent in the gas phase.

Adjusted retention time t_r (min) corresponds to distance BC in *Figure 1*
= characteristic time value for the substance;
= the time spent purely in the liquid phase or on the active solid;
= $t_{dr} - t_d$

Similarly:

Gas hold-up V_d (ml)
= the product of t_d and the volumetric flow rate of the carrier gas F, the latter being measured in ml/min.

Retention volume (uncorrected) V_{dr} (ml)
= product of t_{dr} and F

Adjusted retention volume V_r (ml)
= product of t_r and F, or difference $V_{dr} - V_d$

6

Due to the compressibility of the carrier gas and the influence of the gas laws, the volumes need to be further corrected and related to defined conditions.

The application of the pressure correction (using Martin's factor and taking into account the atmospheric pressure at which the volumetric flow rate F is measured) is indicated by a superscript p on the volume symbol.

The retention volume is influenced by the temperature at which it is measured, and to an even greater extent by the column temperature. The latter influence is far greater than the flowmeter temperature. Thus a further correction taking this temperature into account is unavoidable, and is indicated by a superscript T on the volume symbol. The symbol V_r^{pT} therefore indicates that both temperature and pressure influences have been taken into account. If we remember that the column temperature plays a part which is far greater than that of the temperature at which the volume is measured, we shall find it easier to correct the volume to the column temperature T (in degrees Kelvin), instead of to 0°C, as has been the custom.

Besides the specific retention volume V_g (obtained by relating V_r^{pT} to 1g of liquid phase), the partition coefficient K, and the retention index I (see also Volume III), *relative* values are also used for the qualitative evaluation of chromatograms. We must, however, take great care to distinguish between relative values which are derived from the retention time (uncorrected) t_{dr} and those derived from the adjusted retention time t_r.

The former are called relative retention times and are given the symbol α, the latter are called retentions and are given the symbol r. Of these also, detailed examples will be found in Volume III.

Quantitative evaluation is carried out by, among other things, determining the peak areas from the

peak height and the
peak width at half height. Naturally there are several methods for doing this, which are dealt with in more detail in Section 3.

The analytical result also depends upon the quality of the detector, which is affected by

noise (in millivolts) and
drift (volts per hour) together with specific
sensitivity and other factors which are dealt with in the appropriate sections; it also depends upon the quality and operating conditions of the column, which are given by the terms

column performance n'
separation factor β
number of theoretical plates n
selectivity s
height equivalent to a theoretical plate HETP
resolution ϑ

7

and other special expressions, which are dealt with in more detail in the section on columns.

History of Gas Chromatography

Gas chromatography is assumed to be a very recent technique; nevertheless, if we do not take the previously given definition too strictly, we find that this technique was first used for preparative purposes (for the preparation of pure ethyl alcohol!) by a Strassburg surgeon called Brunschwig in 1512. Bayer and Born[1] have followed the almost 450 year old directions and confirmed their validity.

The work of James and Martin[2] was, however, without doubt the decisive step for the unexpectedly rapid and widespread development of the methods of gas–liquid partition chromatography, which is the chief method used today. With their now world-famous work on the rapid qualitative and quantitative analysis of very small quantities of a mixture of the fatty acids from formic to lauric acid inclusive in 1952, they initiated one of the most remarkable recent developments in analytical chemistry. Of course, even as early as 1941 it had been suggested by Martin and Synge[3] that it should be possible to separate a mixture of substances on the basis of partition and development in the gaseous state. In 1943 Damköhler and Theile[4, 5] used partition gas chromatography, apparently without realizing it, when they separated methanol/ethanol and benzene/cyclohexane mixtures by passing them through 4 m columns containing clay as solid support and glycerine as liquid phase, with H_2 or N_2 as carrier gas. Without any doubt the oldest source for adsorption gas chromatography by the elution technique, apart from the data given by Ramsay in 1905 [26], is P. Schuftan's book *Die Technische Gasanalyse* published by S. Hirzel in Leipzig in 1931, which was followed by the work of Hesse *et al.* in 1941[6], Hesse and Tschachotin in 1942[7], Damköhler and Theile in 1943[4, 5], Wicke in 1946[7], Cremer in 1951[9], Janak in 1953/54[10, 11], and Patton *et al.* in 1955[12].

Not quite so far back come the first works on adsorption gas chromatography by the displacement technique. The development of this technique apparently began with Turner in 1943[13], Claesson in 1946[14, 15, 16], Phillips in 1949[17], Turkel'taub in 1950[18], and James and Phillips in 1953/54[19, 20, 21]. It is now established as a convenient and proven process especially for preparative purposes.

Work on adsorption gas chromatography by the frontal analysis method was chiefly described by Phillips in 1953/54[19, 20, 21]. In 1959 Janak[27] extended the scope of gas chromatography to include the direct analysis of non-volatile materials, after he had been able to show by systematic experiments that such substances could be thermally decomposed in a reproducible manner to give substances which were capable of undergoing gas chromatography.

Scarcely a decade after the pioneer work of James and Martin a reasonably complete survey of the literature of gas chromatography can only be made with the aid of mechanical or electronic devices (punched card systems, etc.), and almost every year there are symposia, including several national and one international, on the theory, techniques and application of gas chromatography, to say nothing of the numerous practical courses which are held in a number of countries. Although the first records of gas chromatography go back almost 450 years, its true history started scarcely ten years ago. It should not be forgotten that Tswett[22] carried out the first chromatographic experiments by the liquid–solid method in 1902, which were only taken up again 25 years later by Kuhn, Winterstein and Lederer[23], when they separated plant pigments by the liquid–solid method (elution technique). It was not until another ten years had passed that Tiselius in 1940/43[24, 25] and Claesson in 1946[14] described the two other techniques of liquid–solid chromatography, namely frontal analysis and displacement analysis. Gas chromatography is thus the most recent branch of chromatography.

References

1. BAYER, E., *Gaschromatographie*, Springer-Verlag; Berlin, 1959.
2. MARTIN, A. J. P. and JAMES, A. T., *Biochem. J.*, 1952, **50**, 679.
3. MARTIN, A. J. P. and SYNGE, R. L. M., *Biochem. J.*, 1941, **35**, 1358.
4. DAMKÖHLER, G. and THEILE, H. *Angew. Chem.*, 1943, **56**, 353.
5. DAMKÖHLER, G. and THEILE, H., *Beih. Z̧. Ver. dtsch. Chem.*, 1943, No. 49.
6. HESSE, G., EILBRACHT, H. and REICHENEDER, F., *Liebigs Ann.*, 1941, **546**, 233.
7. HESSE, G. and TSCHACHOTIN, B., *Naturwiss.* 1942, **30**, 387.
8. WICKE, E., *Angew. Chem.*, 1947, **19**, 15.
9. CREMER, E. and PRIOR, F., *Z. Electrochem.*, 1951, **55**, 66.
10. JANAK, J. and RUSEK, M., *Chem. Listy*, 1953, **47**, 1190.
11. — *Chem. Listy*, 1954, **48**, 207, 397.
12. PATTON, H. W., LEWIS, J. S. and KAYE, W., *Analyt. Chem.*, 1955, **27**, 170.
13. TURNER, N. C., *Nat. Petrol. News*, 1943, **35**, R 234.
14. CLAESSON, S., *Ark. Kemi, Min. Geol.*, 1946, **A23**, No. 1, 133.
15. — *Ark. Kemi, Min. Geol.*, 1946, **A24**, No. 7, 7.
16. — *Disc. Faraday Soc.*, 1949, No. 7, 34.
17. PHILLIPS, C. S. G., *Disc. Faraday Soc.*, 1949, No. 7, 241.
18. TURKEL'TAUB, N. M., *Zhur. Anal. Khim.*, 1950, **5**, 200.
19. JAMES, D. H. and PHILLIPS, C. S. G., *J. Chem. Soc.*, 1953, 1600.
20. — *J. Chem. Soc.*, 1954, 1066.
21. GRIFFITHS, J. H. and PHILLIPS, C. S. G., *J, Chem. Soc.*, 1954, 3446.
22. TSWETT, M., *Ber. dtsch. bot. Ges.*, 1906, **24**, 316, 384.
23. KUHN, R., WINTERSTEIN, A. and LEDERER, E., *Hoppe-Seyl Z.*, 1931, **197**, 141.
24. TISELIUS, A., *Ark. Kemi, Min. Geol.*, 1940, **14B**, No 22, 5.
25. — *Ark. Kemi, Min. Geol.*, 1943, **16A**, No. 18, 11.
26. RAMSAY, W., *Proc. Roy Soc.*, 1905, **A76**, 111.
27. JANAK, J., paper presented at the 7th gas chromatography conference of the Czechoslovakian Study Group in Prague, September 2-3, 1959.

Further Literature
HARDY, C. J. and POLLARD, F. H., Review of gas-liquid chromatography (up to 1959). *J. Chromatog.*, 1959, **2**, 1.

1. THEORY OF GAS CHROMATOGRAPHY

The following is a brief summary of the theory of gas chromatography. The most important laws and relationships for the analyst are given, together with the relationships between them.

CHROMATOGRAPHIC methods are physical or physico-chemical methods for the separation of materials. The components to be separated are distributed between two non-miscible phases in a molecular, i.e. dissolved, form. One of the two phases is normally stationary, and has a large surface area relative to its volume. The other phase moves over the stationary phase in a finely divided state. The mobile phase transports the components to be separated.

Of interest to the practician are the methods used and conditions required for the optimum solution of a given separating problem. In order to arrange for optimum working conditions he must know the way in which very different factors work together and their functional relationship. It is for this reason alone that the theoretical relationships are presented.

The theories of gas chromatography have kept pace with the development of its practical applications. There are basically two different aspects from which it may be regarded: the thermodynamic and the kinetic. Recently[1] a purely statistical-mathematical treatment of the problems of gas chromatography has been published, which leads to practically the same conclusions as the purely thermodynamic method. The entire process can also be represented by an electrical analogue of the column, which served for the basis of Golay's[2] mathematical theory of gas chromatography. However, we require a physical theory for the methods of gas chromatography which will give us a quantitative relationship between column efficiency and the experimental parameters (temperature, type of carrier gas, flow rate, packing, and time required for an analysis).

It will be preferable to use the kinetic approach. We shall follow individual molecules as they move through the column. Cremer[3] and Müller used an original analogy in order to make the process of separation easier to understand. The column is regarded as a river which links the towns A and D, the direction of the current being from A to D. 'Let us imagine that on the river there is a series of vessels which are propelled solely by the current. Along the banks of the river there are landing-stages. Should a vessel touch one of these landing-stages it will be held there for a greater or lesser time. Those vessels which make only a brief stop will be the first to reach the mouth of the river (town D) and those which make long stops will arrive at a correspondingly later time. Those vessels which have the same average rest periods will arrive at the same time. We have thus three requirements for the chromatographic process. There must be:

1. An adsorbent (liquid phase)
2. Substances which are adsorbed on it (or dissolved in it)

10

3. A current to carry the substances over the adsorbent (column packing). The current may be a stream of liquid or of gas.'

The analogy used by Cremer and Müller can be extended still further. Let us suppose that the stream carries not merely vessels but boats of different sizes and manned by different types of people. Those boats which are above a certain size cannot put in at every landing-stage, and so the vessels do not all make the same number of stops. Nor do the passengers all spend the same amount of time at the stopping-points; one group of passengers is not at all interested in visiting them, and consequently makes no stops. A second group consists of people who stop only a short while. The third group visits every stopping place and spends a long time at each.

Let us consider the rates of travel of the three groups. As the first group makes no stops and thus experiences no delay it will arrive at D in the shortest possible time. But even though all the boats started from A at the same time, the vessels in Group 1 would still not all arrive at the same time at D.

Only a very few boats spend the whole of their travelling time in the middle of the river where the current speed is greatest. Some of the boats are more or less near the bank, where they travel more slowly because of the lesser rate of flow, and some boats suffer further delay as a consequence of mutual obstruction.

Thus at D a small proportion of the boats arrive at a time $t_d - \frac{1}{2}\Delta$, a larger number at a time t_d, and another small number at a time $t_d + \frac{1}{2}\Delta$.

Provided that sufficient boats of Group 1 started out from A, the number of boats arriving at D plotted against their time of arrival would have the form of a Gaussian function.

We may refer to this dispersion process as diffusion. The longer the stretch of river chosen or the wider the river relative to the size of the boats, the further apart the boats would diffuse.

Even if all the boats were transported at the same rate by the current in the middle of the river, the boats in Group 2, which break their journey for a rest, would arrive later at D. These boats would also be further 'diffused apart', because to the statistical irregularities of river travel would be added statistical irregularities in the length of the rest period. If number and time of arrival are plotted against one another in *Figure 2* we get another Gaussian function, i.e. a peak, this time with its maximum at $t_d + \tau$; τ may be regarded as the average rest period for a given group.

By methods analogous to those previously used we can establish that all the boats of the third group will, on the average, arrive at a time $t_d + n\tau$ at D, where n is the number of times they have stopped and τ is the number of minutes spent at each stop.

In what way do the three groups differ from one another? The picture one would get at D would show that each group takes a different time to travel the distance. As the distance does not vary, it appears as if the three groups must have different rates of travel. But we know that the rate of travel for the three groups is the same and is in fact equal to the rate of

11

flow of the river, and that the difference between the groups lies in the number and duration of the stops they make.

Thus the characteristic values for the groups are not the total travelling times but the times of rest. The times of rest may be obtained from the total travelling time t_{dr} by subtracting the minimum travel time t_d, which depends on the distance and nature of the journey.

$$t_r = t_{dr} - t_d \qquad \qquad ....(1)$$

If, due to external circumstances, the flow rate of the river changes, then the total travelling time will also change. It is therefore more correct to use the volume of water flowing in the same time as our basic unit. In this way variations in the rate of flow are allowed for.

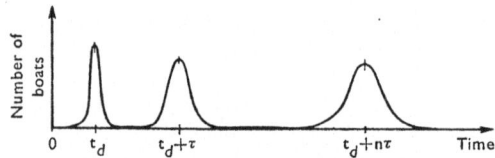

Figure 2. Explanation of the processes of separation and diffusion

Thus time multiplied by flow equals volume:

$$t \cdot F = V \qquad \qquad ....(2)$$

At this point we shall leave our model. It has shown how components may be separated, even though they are travelling in the same direction and at the same speed. Their characteristic differences are derived from the number and duration of their rest periods, and depend on the characteristic differences between the components and the specific properties of the 'landing-stages'.

Point A is thus the sample injection point, and point D is the detector. The river with its banks is the carrier gas and the stationary phase. The rest periods represent the retention of the substance by the liquid phase or the active solid.

In practice, conditions are naturally more complicated. The transporting medium is a gas. It is compressible. The column sets up a resistance to the flow of the gas. The gas pressure is higher at the column inlet than at the column outlet. The rate of travel of the carrier gas and therefore all the time and volume data are greatly dependent on the operating conditions.

In the second place the individual molecules have very different distances to travel, because the paths around the particles of the column packing vary considerably.

The characteristic values given in the following section are therefore average values.

Definition and Derivation of the Characteristic Values for Gas Chromatography

Let us follow the progress of a molecule from the point where it is in the gas phase. It travels at a rate $u = L/t_{dr}$ in the direction of flow of the carrier gas along the entire length L of the column. Now and again it collides with the phase boundary, and each time is retained there for a time τ. Let the number of successful transitions from the mobile to the stationary phase be n. If the molecule made no contact with the stationary phase, the time needed to pass through the column would be t_d. But the total time of travel is:

$$t_{dr} = t_d + n\tau$$

assuming that the molecule passes over into the stationary phase on n occasions and is retained there each time for a period of τ minutes.

In conformity with equation 1 we replace $n \cdot \tau$ with t_r; we then have

$$t_{dr} = t_d + t_r$$

The time characteristic for the retention of the material in the liquid phase of the column packing is called the retention time, and is given in minutes.

Two further time values need to be distinguished: the time the substance needs just to pass through the column, or in other words the time spent in the gas phase in the column, and the sum of these two time values. Thus we may define:

t_{dr} (min) = retention time (uncorrected) (total time needed to pass through the column)
t_r (min) = adjusted retention time (retention in liquid phase)
t_d (min) = hold-up time (retention in gas phase)

As noted previously, it is more expedient to give, not the time, but the volume of gas flowing through the column in this time as the characteristic value. They are named analogously:

V_{dr} (ml) = retention volume (uncorrected)
V_r (ml) = adjusted retention volume
V_d (ml) = hold-up volume, which is equal to the volume of gas in the column

These values are given in ml and are obtained by multiplying the time by the volumetric flow rate (in ml/min). The appropriate gas laws require that temperature and pressure should be taken into account in order to obtain comparable and defined volumes. The following values affect the volume:

1. Column inlet pressure p_i (mm Hg at 0°C).
2. Column outlet pressure p_o (mm Hg at 0°C).
3. Temperature at which the gas flow is measured (T_m) (degrees Kelvin) and the column temperature (T_s) (degrees Kelvin).

13

These factors are involved in the conversion of time values to volumes; correction for them is indicated by superscripts to the volume symbol. We shall use the pressure correction factor given by Martin and James[4-7]. This, however, calls for a few special comments.

In recent years the pressure correction factor has been published in forms which give rise to misunderstandings and errors. Thus there is in existence a whole series of publications in which the factor is given in an incorrect form. In order that the user shall be in a position to distinguish the correct from the incorrect forms, we shall examine the significance of the pressure correction factor more closely.

Because of the compressibility of the carrier gas, for a given pressure drop along the column $p_i - p_o$ the quantity of carrier gas issuing from the column is greater than would be expected from a linear relationship to the pressure gradient. Because of this the flow rate, which can only correctly be measured at the column outlet, increases with increasing pressure drop to a greater extent than would correspond to a directly proportional increase in pressure drop. Thus the retention volume increases with increasing pressure drop. But if the retention volume is to be a specific value with which qualitative identification of a given substance can be carried out, then it must remain constant. If the carrier gas was not compressible, then the flow rate would increase linearly with increasing pressure drop, and as the retention time would decrease linearly the product $t_r . F$ would always be constant, so that the retention volume would be a constant, i.e. a value independent of the pressure.

However, as the value for the flow rate is greater than would be expected from its proportionality to the pressure drop, the numerical value of the correction factor must be less than 1, and as the pressure drop increases it will become less and less. A table enabling the pressure correction factor to be read off for every value of p_i and p_o will be found in Volume III.

The above requirements are met only by the form

$$f = \frac{3\left[\left(\frac{p_i}{p_o}\right)^2 - 1\right]}{2\left[\left(\frac{p_i}{p_o}\right)^3 - 1\right]} \text{ for } V_r^{\text{Pressure corrected}} = V_r . f$$

The factor f can also be used in the form

$$f = \frac{3\left[\frac{p_i}{p_o} + 1\right]}{2\left[\left(\frac{p_i}{p_o}\right)^2 + \frac{p_i}{p_o} + 1\right]} \quad ....(3)$$

which simplifies the calculation.

Absolutely correct values are still not obtained from the Martin factor in the above form if V_r values for relatively short columns (less than 3 m) with high pressure drop are to be corrected. While the Martin factor works very well for capillary chromatography, a factor f', corrected in the following way, gives even better results which are also valid at a relatively high pressure drop

$$f' = \frac{3\left[\left(\frac{p_i}{p_o}\right)^2 + \frac{p_i}{p_o} + 1\right]}{2\left[\left(\frac{p_i}{p_o}\right)^3 + \left(\frac{p_i}{p_o}\right)^2 + \frac{p_i}{p_o} + 1\right]}$$

Care should be taken that the values for p_i and p_o should always be given as absolute pressures:

p_i in atmospheres absolute or mm Hg at 0°C

p_o in atmospheres absolute or mm Hg at 0°C

For example: if the pressure on the manometer at the column inlet is read off as 1 atm gauge pressure this is 2 absolute atmospheres, and we can say $p_i/p_o = 2/1$

For this value

$$f = 0.64286$$
$$f' = 0.7000$$

if p_o is exactly 760 mm Hg = 1 absolute atmosphere.

f' is especially to be used for rapid analyses, but the validity of both correction factors is limited.

The fact that the pressure has been corrected is shown by a superscript p:

$$t_r \cdot f \cdot F_m = V_r^p$$

The correction is, of course, only valid if p_o is taken not merely as the general atmospheric pressure but as the pressure actually prevailing at the measuring instrument during the measurement of the flow rate F_m (ml/min). (Thus pressure head, water vapour partial pressure, etc., must be taken into account.)

The retention volume is only exactly defined when the temperature at which it arises and is measured is also taken into consideration. It might be thought that a gas volume should always be reduced to normal conditions, i.e. 760 mm and 0°C. But the retention volume depends to an extraordinarily great extent on the column temperature, since the retention time is influenced by the temperature of separation, and thus it can be seen to be better to correct the retention volume to the column temperature. Indeed, in view of the fact that it is impossible to compare retention values without also giving the column conditions, the requirement that the retention volume should be reduced to 0°C loses importance. Finally, the advantages of giving the volume at column temperature appear also in the further evaluation of the retention volume for the conversion to the partition coefficient. For this reason we shall always give both the retention volume and also the partition coefficient (see below) reduced to the column temperature, because the influence of the latter on the values is so strong as to be dominant. The exception is the specific retention volume, which is reduced to 273° Kelvin. This is achieved in practice by correcting the flow rate F_m, which is measured at the temperature of the measuring instrument T_m, to the column temperature T_s by the equation

$$F_m \cdot \frac{T_s}{T_m} = F_s \qquad \qquad ....(4)$$

where T_s and T_m are the column and measuring instrument temperatures in degrees Kelvin, and F_m is measured at the column outlet at a pressure p_0.

The fact that the value has been corrected to the separation temperature is shown by a superscript T, or in special cases (where concrete values are referred to) by the temperature in degrees Kelvin, also written superscript:

$$t_r \cdot f \cdot F_m \cdot \frac{T_s}{T_m} = V_r^{pT} \quad (ml) \qquad \qquad ....(5)$$

$$V_r \text{ corrected to } 80°C = 350°K: \ V_r^{p350}$$

This form may appear complicated, but it is necessary in view of the strong temperature dependence of the retention values. Correspondingly:

$$\text{Corrected total retention volume: } V_{dr}^{pT} \quad (ml) \qquad ....(6)$$

$$\text{Gas volume of column: } V_d^{pT} = V_G \quad (ml) \qquad ....(7)$$

For a given liquid phase or active solid the corrected retention volume V_r^{pT} is also dependent on the quantity of effective stationary phase, i.e. on the quantity of liquid phase or active solid which comes into contact with the gas phase.

Under normal conditions the retention time increases linearly with increasing quantity of liquid phase. This variable may be eliminated by reducing the retention values to 1 g of effective stationary phase; by converting this value to 273°K we obtain the specific retention volume V_g. The reduction to 273°K is to a certain extent a luxury, since it means that the V_g values cannot be quoted without also quoting the column temperature and the liquid phase. But it is not in the interests of international comparability of values to ignore the IUPAC recommendations.

The value of V_g may thus be obtained from the equation

$$\frac{V_r^{pT} \cdot 273}{W_L \cdot T_s} = V_g \ (ml/g) \qquad \qquad ....(8)$$

W_L = grams of liquid phase

T_s = temperature of separation in degrees Kelvin.

V_r^{pT} = retention volume corrected for temperature and pressure.

V_g, the specific retention volume, is one of the fundamental values of gas chromatography.

At this point it should be noted that with a normal commercial instrument it is not possible to determine exact values for V_g or the values of K derived therefrom, because neither the manometer and flowmeter nor the air thermostat used in such instruments work to a sufficient degree of accuracy. Even if a constant temperature can be simulated in the air thermostat by the use of a mercury thermometer which is as insensitive as possible, there still exists within most commercial thermostats so great a temperature gradient that measured and corrected values of the necessary degree of reliability and accuracy cannot be obtained.

For a correctly determined V_g value all the apparatus values such as the column length, diameter and pressure drop are eliminated. However, the restriction mentioned earlier that the essential quantity of liquid phase or active solid was the *effective* quantity cannot be exactly defined without going a little further. A diminution of the effective quantity of liquid phase or adsorption surface due to the particular structure of the carrier material can influence the retention values.

The same is true of faults in the apparatus (dead volume, incorrect column packing, etc., derangements in the distribution of substance between mobile and stationary phase) and systematic errors, which occur especially during sample injection.

If the injection is carried out too slowly, or if the substance—where this is originally in a liquid or solid form—is not vaporized immediately and completely, or if the load capacity of the column is exceeded, the separating capacity and column performance will always be strongly and negatively affected. The retention values may thus only be regarded as exactly reproducible values if all the above-mentioned errors are avoided; the influence of the quantity of substance injected does not become nil only on extrapolating to theoretically ideal conditions, i.e. where in the case of injection one reduces to nil g of injected substance, but when the quantity injected does not exceed the capacity of a theoretical plate. This point will be returned to later (see Littlewood, Phillips and Price[8]).

Retention Volume and Partition Coefficient

The partition coefficient gives the ratio of the distribution of the substance between the liquid phase and the gas phase.

Definition: $K = \dfrac{\text{grams of substance in 1 ml liquid phase}}{\text{grams of substance in 1 ml carrier gas}}$

The value of K is high when most of a substance is retained in the liquid phase. This means that the substance only moves slowly down the column because only a very small fraction will be in the carrier gas and it is only this fraction which is moved on. But such a substance has a high retention volume. There is thus a simple relationship between retention volume and partition coefficient:

$$V_r^{pT} = K \cdot V_{L(T_s)}$$

$$K = \frac{V_r^{pT}}{V_{L(T_s)}} \qquad \ldots (9)$$

T_s = Temperature of the column in degrees Kelvin

$V_{L(T_s)}$ = Volume of liquid phase in the column at T_s degrees Kelvin.

Further, we may say

$$K = V_g \frac{T_s \cdot \rho L(T_s)}{273} \qquad \ldots (10)$$

$pL(Ts)$ = density of the liquid phase at a column temperature of $T_s(°K)$.

Also
$$V_g = \frac{K \cdot 273}{T_s \cdot pL(Ts)} \qquad \text{....(11)}$$

K gives us a further fundamental value which may easily be obtained from gas chromatographic values and at the same time can be calculated free from all apparatus variables. It is, however, necessary to calculate the density of the liquid phase at the column temperature T_s, i.e. the value $pL(Ts)$, from the weight of liquid phase in the column W_L and the density of the liquid phase at 20°C $pL(293)$.

This may be carried out by the following approximation:

$$p\,L(Ts) = p\,L(293) \cdot (1 + (T_s - 273) \cdot 10^{-3})(\text{g/ml}) \qquad \text{....(12)}$$

The value $V_{L(Ts)}$ (volume of liquid phase at column temperature) may be obtained from the familiar equation

$$V_{L(Ts)} = \frac{W_L}{pL(Ts)} \quad (\text{ml})$$

Retention Values and the Liquid Phase in Isothermal Gas Chromatography

Different types of separation often require different liquid phases. While, for example, a mixture of saturated *n*-alkanes can be separated on almost any liquid phase, the separation of the aromatic hydrocarbons into all their different isomers requires a liquid phase which is completely specific. Because of this a whole series of further experimental factors have to be taken into account. All these correlations are dealt with in the section on columns. Here we shall only deal with the simplest of the formal relationships.

It has been found experimentally that within a homologous series there exists a linear relationship between the logarithm of the retention value and the molecular weight or the number of carbon atoms in the homologue:

$$\log V_g = k \cdot n \qquad \text{....(13)}$$

k = constant, dependent on the type of homologous series, the type of liquid phase and the temperature.

n = molecular weight or carbon number of the members of the homologous series (e.g. the *n*-fatty acids).

This discovery was made by James and Martin[4] and it has been confirmed by numerous later investigations, among them those of Ray[9]. Of course the first members of the series, especially in the case of the polar compounds (acids, alcohols, etc.) deviate slightly from the straight line relationship.

Of especial importance for the qualitative analytical application of gas chromatography is the linear relationship, first discovered by Martin *et al.*[6], between the retention values of a homologous series in a column A and the retention values of the same series in a different (i.e. filled with a different liquid phase) column B:

$$V_{gA} = k'' \cdot V_{gB} + C \qquad \text{....(14)}$$

k'' has a different value for each homologous series; V_{gA} is the specific retention volume on column A; C = constant.

The individual classes of compounds may be recognized by the differing gradients of the curves. It is especially instructive to plot the logarithms of the retention values for one column against those for a different column.

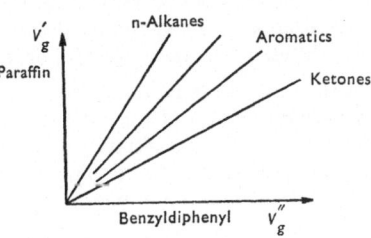

Figure 3. Illustration of equation 14. Constant C can take the value 0

This method, which was first proposed by Pierotti et al.[10], may be used directly for qualitative identification.

In most cases a series of parallel straight lines is obtained, whose position and separation are characteristic for the functional groups of the various homologous series.

$$\log V_{gA} = k''' . \log V_{gB} + C_{A,\,B} \ldots \qquad \ldots(15)$$

k''' is the same for all homologous series; $C_{A,B} \ldots$ is different for each series.

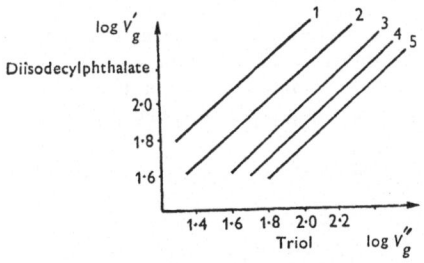

Figure 4. Illustration of equation 15.
Triol = triethylene glycol.
1 = n-alkanes.
2 = ketones.
3 = tertiary.
4 = secondary.
5 = primary alcohols.
(from Pierotti *et al., J. Amer. Chem. Soc.,* 1956, **78**, 2989)

The retention values are quantitatively influenced by the Debye, London, Keesom and chemical bonding forces between the liquid phase and the substance (see p. 53).

Height Equivalent to a Theoretical Plate

The analogy of the gas chromatographic process used earlier—a river with boats drifting down it—showed clearly how the substance zones, which at the start were quite narrow, underwent a spreading effect during the course of the process. A zone of substance moving through the column with the carrier gas is subject to spreading influences, due to

(1) longitudinal diffusion in the carrier gas along the column axis;

(2) formation of a flow profile corresponding to the viscosity and flow rate of the carrier gas;

(3) retardation of the substance as it passes into the stationary gas layers surrounding the large surface areas of the solid support particles;

(4) retardation as the substance passes into the liquid phase and is retained there (diffusion through a greater or lesser thickness of liquid film).

As the mutual interaction of the carrier gas flow rate, the form, amount and packing density of the solid support, the pressure gradient along the column due to this, the viscosity of the carrier gas, the average viscosity and film thickness of the liquid phase—and also the form of the pores and channels filled with liquid phase—and finally the ratio of the times spent in the liquid phase and in the carrier gas, since the mutual interaction of all these factors is so complex it is quite understandable that a great deal of theoretical and experimental work has been put into the quantitative elucidation of these relationships.

The ratio between the spreading of the substance band (measured as a time value) and the retention time in relation to the column length is called the height equivalent to a theoretical plate (*HETP*).

$$HETP = 180 \cdot 5 \cdot \left(\frac{b_{\frac{1}{2}}}{t_{dr}}\right)^2 \cdot L \quad (\text{mm}) \qquad \dots (16)$$

$b_{\frac{1}{2}}$ = peak width at half height measured in mm or sec.

t_{dr} = retention time (uncorrected) measured in mm or sec.

L = length of effective column packing measured in m.

The smaller the height equivalent to a theoretical plate, the easier it is to carry out a separation in a short time and with a short column.

The following simple relationship exists between the number of theoretical plates in a column and the height equivalent to a theoretical plate:

$$n = \frac{1,000 \cdot L}{HETP} \qquad \dots (17)$$

L = column length in m; *HETP* is given in mm.

It was Synge and Martin[15] who first introduced the theoretical plate concept for the treatment of the chromatographic process.

Later, Mayer and Tompkins[16] extended Martin's theoretical plate theory and calculated the number of theoretical plates necessary to achieve a given purity for the substances separated. The theory also enables the form of the peaks at the column outlet to be calculated. Glueckauf[17] corrected the discontinuous theory of Mayer and Tompkins and replaced it with one in which mass transfer between the phases took place continuously. The difference between the two theories is that for the same number of theoretical plates the Gaussian curves have different widths.

Finally, van Deemter, Zuiderweg and Klinkenberg[18] extended the theories of Glueckauf and other workers to give the so-called rate theory, which gave a correct qualitative description of the broadening of the Gauss

Height Equivalent to a Theoretical Plate

curve due to the non-ideal chromatographic behaviour of the packed column, and gave a partial quantitative explanation.

Van Deemter *et al.* obtained a relationship which when written in the form used by Keulemans and Kwantes[19] is as follows:

$$HETP = A + B/u + C \cdot u \qquad \qquad \text{....(18)}$$

A, *B* and *C* are constants whose values are dependent on the operating conditions of the column and similar parameters;
u is the average linear carrier gas flow rate (cm/sec).

Meanwhile corrections and extensions have also been applied to this equation, so that finally according to the researches of Jones and Kieselbach[26] it reads:

$$HETP = A + B/u_0 + Cu_0 + Du_0 + Eu_0$$

or written out in full:

$$HETP = A + B' \frac{D_{go}}{u_0} + C' \frac{k}{(1+k)^2} \cdot \frac{d_l^2 \cdot u_0 \cdot 2p_0}{D_l(p_i + p_o)} + D' \frac{k^2}{(k+1)^2} \cdot \frac{d_m^2 \cdot u_0}{D_{go}}$$

$$+ E' \frac{d_s^2 \cdot u_0}{(1+k)^2 \cdot D_{go}} \qquad \qquad \text{....(19)}$$

u_0 (cm/sec)	= linear gas velocity at outlet of column packing
A (cm)	= velocity-independent constant of random flow pattern in a packed column
B'	= coefficient of axial diffusion in the gas phase
C' (sec)	= coefficient of resistance to mass transfer in liquid phase
D' (cm²)	= coefficient of resistance to mass transfer in flowing gas phase
D_{go} (cm²/sec)	= coefficient of diffusion of the substance in the carrier gas
D_l (cm²/sec)	= coefficient of diffusion of the sample in the liquid phase
d_l (cm)	= effective film thickness of the liquid phase
d_m (cm)	= effective diffusion path length in the flowing gas
d_s (cm)	= effective diffusion path length in the stationary gas
E'	= coefficient of relatively stationary gas within column packing
p_i (atm)	= gas inlet pressure
p_o (atm)	= gas outlet pressure
$k = t_r/t_m$	= ratio of adjusted retention time to hold-up time

The manner in which the factors *A-E* are affected by the retention, or the values of *k*, is shown in the following table for air ($k = 1$), butane ($k = 2$), and cyclohexane ($k = 3$) on a 1 m column (see p. 22).

The *HETP* can thus be influenced by the type of liquid phase—it should have as low a viscosity as possible and be spread over the solid support in a very thin film—by the granule size of the solid support (it should be between 0·125 and 0·150 mm) and by the type of packing.

Even when these values have been settled, the column efficiency can still be affected by the gas flow. There is an optimum value for this, for the above equation $HETP = f(u_o)$ is in the form of the equation for a hyperbola with

	A	B	C	D	E
Air	0·008	0·026	0·0	0·0	0·026
Butane	0·008	0·021	0·014	0·005	0·003
Cyclohexane	0·008	0·016	0·003	0·013	0·0

a minimum for $HETP$ at an optimum value of u_o. As the carrier gas is compressible, the optimum gas rate can only be maintained over a given section of the column. The important thing is to make this section as long as possible. Practical hints which are of importance in the preparation of a good column will be found in the section on columns.

The fundamental validity (even under extreme conditions) of the equation of van Deemter *et al.* has been confirmed by Brennan and Kembal[20]. For packed columns an $HETP$ value of 0·3mm may be obtained.

Separation Formula for Gas Chromatographic Methods

On the basis of the relationships already established by Consden, Gordon and Martin[12] and James and Martin[4], Herington[11] derived the fundamental separation formula for gas chromatography. It presents the relationship between the interaction of the properties of the substance and the stationary phase in the column, and indicates the separation that can be achieved. The equation is:

$$\log \frac{V_{g2}}{V_{g1}} = \log \frac{p_1^o}{p_2^o} + \log \frac{\gamma_{13}^o}{\gamma_{23}^o} \qquad \dots (20)$$

V_{g1} = specific retention volume of substance 1
p_1^o = vapour pressure of substance 1 under normal conditions
γ_{13}^o = activity coefficient of substance 1 at infinite dilution in stationary phase 3, reduced to normal conditions.

This means that substances can be separated even if they have practically identical vapour pressures, provided that a specific liquid phase can be found. It further implies that for gas chromatography there is no such thing as an inseparable azeotrope.

The equation was derived on the basis of mixed phase thermodynamics; detailed methods for its derivation have been given by Röck[13] among others.

Derivation of the Separation Formula

By definition, the partition coefficient K is the ratio of the quantity of substance dissolved in 1 ml liquid phase to the quantity of gaseous substance in 1 ml carrier gas.

22

For ideal solutions K is completely independent of concentration, and this is found in practice in the case of very dilute solutions.

For ideal solutions, Raoult's law requires that the vapour pressure of a volatile component over its solution in a non-volatile solvent should be dependent on the concentration and the vapour pressure of the pure component, according to the equation:

$$p = x \cdot p^\circ \qquad \dots(21)$$

where x = mole fraction of the component in the solution.

As, however, two mutually soluble substances are only very rarely able to form ideal solutions, equation 21 hardly applies.

We therefore need to introduce a factor γ, which is dependent on the nature of the substance and which can be inserted in equation 21. This factor is called the activity coefficient.

$$p = \gamma \cdot x \cdot p^\circ \qquad \dots(22)$$

The partial pressure p can equally well be expressed in terms of the total pressure P over the solution by using

$$p = y \cdot P \qquad \dots(23)$$

where y = mole fraction of the component in the gas phase.

We can thus say

$$yP = x \cdot \gamma \cdot p^\circ \qquad \dots(24)$$

For dilute solutions the ratio x/y is by definition connected with the partition coefficient K by the equation:

$$K = \frac{x}{y} \cdot \frac{N_L}{N_G} \qquad \dots(25)$$

where N_L = moles liquid phase per ml and N_G = moles carrier gas per ml.

Substituting equation 24 gives

$$K = \frac{P}{\gamma p^\circ} \cdot \frac{N_L}{N_G} \qquad \dots(26)$$

and as $P \cdot V = n \cdot RT$, P/N_G can be replaced by RT:

$$K = \frac{N_L \cdot RT}{\gamma \cdot p^\circ} \qquad \dots(27)$$

Substitution of equation 10 or 11 in equation 27 gives

$$V_r^{pT} = \frac{N_L \cdot RT}{\gamma \cdot p^\circ} \cdot V_L$$

or

$$V_g = \frac{R \cdot 273 \cdot N_L}{\gamma \cdot p^\circ \cdot \rho L(Ts)} \qquad \dots(28)$$

by means of which equations the activity coefficient may be obtained. If the quantity of substance is reduced to a sufficient extent, then instead of the γ-values, which are still concentration-dependent, we obtain the constant γ°-values. Equation 20 may be directly obtained from equation 28 (for details about the determination of γ-values, see Keulemans[14]).

We obtain

$$\frac{V_{g2}}{V_{g1}} = \frac{\gamma_1 p_1^\circ}{\gamma_2 p_2^\circ} \qquad \dots(29)$$

Selectivity, Column Performance and Resolution

These three values provide an adequate means of characterizing a separating column for a given separation. The separation efficiency of a column is composed of the following two mutually independent factors:

the selectivity s

and the column performance n'.

The selectivity is determined almost entirely by the type of liquid phase in gas liquid chromatography and by the type of active solid in gas solid chromatography (it is also affected to a small extent by the operating temperature). It is obtained from the retentions of a given pair of substances which have identical vapour pressures (see also Bayer, *Gas-Chromatographie*, Springer, 1959).

$$s = r_{21} \text{ where } p_1^\circ = p_2^\circ; r_{21} = V_{g2}/V_{g1} \text{ or } t_{r2}/t_{r1}$$

The column performance is defined as the number of theoretical plates in 1 m of column:

$$n' = \frac{n_{\text{total}}}{L} \qquad \qquad ....(30)$$

n_{total} = number of theoretical plates in the column
L = length of column in m

n' may be connected with the previously defined *HETP* by the equation:

$$n' = \frac{1000}{HETP} \qquad \qquad ....(31)$$

Selectivity and column efficiency can be determined when the separation is completed, using only a few gas chromatographic data, or they can be calculated in advance.

The knowledge of these values not only gives a complete picture of the column but also enables extensive predictions and optimization of separation. Further details on the great practical importance of such calculations will be found in the section on columns. Here we shall simply deal briefly with the relationship between selectivity, column performance, and the resolution for a given pair of substances that may be thus achieved.

The fact that such calculations are limited to two-component systems is no disadvantage. If the conditions chosen are sufficient for the separation of the most critical pair of substances then all the other substances will naturally be separated satisfactorily.

The resolution of a pair of substances is defined as the extent to which the peaks of the two substances overlap. This value may be obtained by two measurements. From the line joining the two peak maxima a perpendicular is dropped to the base line through the dip between the two peaks. This gives rise to two lengths ; the longer of the two corresponds to the average height of the two peaks, while the other is shorter by the distance between the minimum and the base line. The ratio of the two distances is defined as the resolution. A resolution of 0·5 thus corresponds to two peaks which overlap in such a way that the minimum between them is at half the height of the peak maxima.

$$\vartheta = \frac{f}{g} \text{ (always less than 1)} \qquad \qquad ....(32)$$

This definition has an advantage over the one used in the English and American literature, in that here the maximum resolution is 1·0 and can only be achieved with an infinitely great number of theoretical plates. This, too, will be dealt with in more detail in the section on columns.

The retention of a pair of substances as a measure for selectivity, the resolution and the number of theoretical plates required are linked in the following equation, which was obtained by Struppe and Kaiser from the equations of Röck[13] and Golay[2]:

$$n = 2 \ln \frac{2}{1-\vartheta} \left(\frac{r\dfrac{k+2}{k}+1}{r-1} \right)^2 \qquad ....(33)$$

n = number of theoretical plates

ϑ = resolution

r = retention = $t_{r2} : t_{r1}$

$k = t_{r2}/t_d$ or V_r^{pT}/V_G (this k corresponds to the k in the previously given equation for $HETP$)

Instead of the number of theoretical plates, we can also obtain the directly measurable separation factor β:

$$\beta = \frac{t_{dr}}{b_{\frac{1}{2}}}$$

$$= 0·429 \sqrt{n}$$

$$= 0·603 \left(\frac{r\dfrac{k+2}{k}+1}{r-1} \right) \Big/ \sqrt{\ln \frac{2}{1-\vartheta}} \qquad ....(34)$$

For many purposes it is simpler to insert the ratio of the uncorrected retention times $t_{dr2}/t_{dr1} = \alpha$ instead of the retention and the k value, because the former values are especially easy to measure from a chromatogram; in this case the relation ($\alpha > 1$) holds good.

$$n = 2 \ln \frac{2}{1-\vartheta} \left(\frac{\alpha+1}{\alpha-1} \right)^2 \qquad ....(35)$$

Retention Values and Temperature of Separation; Retention Temperature

In the section on temperature a detailed study will be given of the temperature factor, which plays an important role in the practice of gas chromatography. Here we shall only give the basic rules.

We may, as is well known, distinguish between isothermal chromatography and chromathermography.

At high temperatures the solubility ratios of the substances in the liquid phase, the viscosity of the carrier gas and the liquid phase, the gas flow rate and the rate of mass transfer between the gas and liquid phases are all considerably different from their values at lower temperatures, and because of this the specific retention volume also varies to a marked degree with temperature. In spite of this it is very easy to formulate the relationships mathematically.

Let us first consider isothermal gas chromatography. We may say

$$\log V_g = \frac{\Delta H_s}{2 \cdot 3 \cdot RT_s} + C \qquad \dots(36)$$

ΔH_s = partial molar heat of vaporization of the dissolved substance from the solution

R = 1,987 cal/degree mol

T_s = column temperature °K

C = constant

ΔH_s itself varies with temperature. These variations are very similar for the individual members of a given homologous series, and thus we may say that the logarithm of the relative specific retention volume has an inversely proportional relationship to the temperature.

$$\log V_g^{\text{rel.}} = \frac{C}{T_s} ; = \log \frac{V_g}{V_{gB}} ; = \log r_B$$

$\log V_g^{\text{rel.}} = \log V_g$ for substance i/$\log V_g$ for reference substance (e.g. for the saturated hydrocarbons pentane is used as reference substance).

C = constant; r_B = retention related to the reference substance.

This relationship was confirmed in particular by the work of Phillips[8] and reminds us of the well-known fact that the logarithm of the vapour pressure varies proportionally with the reciprocal of the temperature. But vapour pressure and V_g are directly connected.

Different, but just as simple are the relationships between column temperature and retention volume of a substance when the temperature of the column varies (linearly) with time, a process known as gas chromathermography.

The relationship between the corrected total retention time V_{dr}^{pT}, the adjusted retention time t_r, and the carrier gas flow F_0 may be expressed

$$V_{dr}^{pT} = \int_0^{t_{dr}} F_0 \cdot dt$$

or

$$\int_0^{t_{dr}} \frac{F_0}{V_{dr}^{pT}} \cdot dt = 1 \qquad \dots(37)$$

F_0 = gas flow at 0°C (ml/min) (according to Habgood and Harris[28])

V_{dr}^{pT} = pressure and temperature corrected total retention volume. If the temperature now increases with time by an amount H (°C/min), we can say

$$H = \frac{dT}{dt} \text{ or } dt = \frac{dT}{H}$$

H = heating rate in °C/min.

From the above equations, it follows that

$$\int_{T_0}^{T_r} \frac{1}{V_{dr}^{pT}} \cdot dT = \frac{H}{F_0} \qquad \ldots (38)$$

T_0 = initial temperature of analysis

T_r = retention temperature, the temperature at which the substance leaves the column.

In order to be able to find the retention temperatures of individual substances for any given heating rate and gas flow, equation 38 is given as a

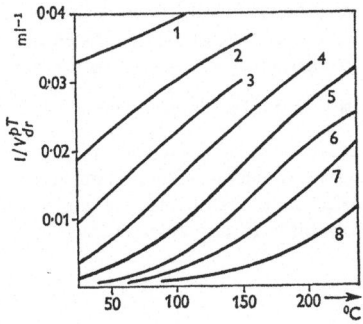

Figure 5. Relationship between $\frac{1}{V_{dr}^{pt}}$
and temperature (°C)T

1 = air.   5 = *n*-hexane.
2 = propane.   6 = *n*-heptane.
3 = *n*-butane. 7 = *n*-octane
4 = *n*-pentane.

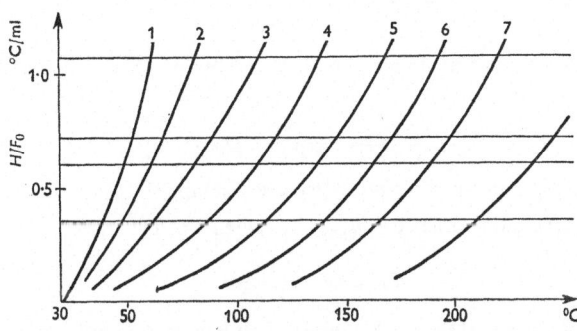

Figure 6. Relationship between $\int_{30}^{T} dT/V_{dr}^{pT} = H/F_0$ (°C/ml) and temperature (from Habgood and Harris, *Analyt. Chem.*, 1960, **32**, 450).

function of the column temperature, or determined graphically from the functions shown in *Figures 5* and *6*.

27

The relation between the time taken for an analysis and the operating conditions will be dealt with in the section on columns.

Number of Theoretical Plates and Peak Form

The length of the column divided by the average *HETP* gives the number of theoretical plates n. The relationship between the number of theoretical plates, the peak form, and the partition coefficient, and also the separation of the substances may be found by continuing the previously mentioned process of calculating the Gauss form of the peaks.

In this way the peaks are first obtained in the form they would have had if no additional spreading due to diffusion had occurred. If the latter factor is taken into account the peaks are obtained in their final form.

The entire column may be regarded as being divided into n theoretical plates of height *HETP*, in each of which (by definition) the partition equilibrium between substance A in the carrier gas and A in the liquid phase is found. The flow of carrier gas ensures that the substance A in the gas phase is continuously carried on to the next plate. For the transport of substance A through all n plates of the column $\frac{t_{dr}}{t_d} V_G = u$. V_G volume units of carrier gas are needed. Each plate is thus cleansed $u \cdot n$ times. The probability for a given molecule of A leaving the column is equal to the probability that during $u \cdot n$ transitions the molecule was in the gas phase n times. This probability P may be calculated according to the laws of probability

$$P_u^n = \frac{u!}{n!\,(u-n)!} \cdot p^n\,(1-p)^{u-n} \qquad \dots (39)$$

p = fraction of total molecules which are in a state of transition between one plate and the next in the gas phase.

By application of Stirling's formula* we obtain, when u, n, and $u-n$ are large:

$$p_u^n = \frac{1}{\sqrt{(2\pi n)}} \cdot \exp\left[- (px)^2 \cdot 2n\right] \qquad \dots (40)$$

for $u_{max.} = n/p$ and $x = -u_{max.}$ (see also Phillips[21]).

P_u^n has a maximum when $u = u_{max.}$, i.e. when $x = 0$.

$$P_{max.} = \frac{1}{\sqrt{(2\pi n)}} \qquad \dots (41)$$

From equation 40 the peak width at half height $b_{\frac{1}{2}}$ may be calculated by using the relationship

$$x = u - u_{max.} \;; \; u_{max.} = n/p; \; p = \frac{V_G}{V_{dr}^{p\,T}} = \frac{t_d}{t_{dr}}$$

* See Sirk, *Mathematik für Naturwissenschaftler und Chemiker*, Theodor Steinkopff, Dresden and Leipzig (1947), p. 139.

We obtain

$$\frac{(px)^2}{2n} = \frac{p^2}{2} \cdot n \left(\frac{t - t_{dr}}{t_d}\right)^2 \qquad \dots(42)$$

where the exponent in equation 40 was first expressed as a function of time. If this value is substituted into equation 40 and the two t values obtained for which $P = \frac{1}{2}P_{\max}$. (i.e. where the separation of the two t values = peak width at half height = $b_{\frac{1}{2}}$), then we obtain

$$b_{\frac{1}{2}} = \frac{2t_d}{p\sqrt{n}} \cdot \sqrt{(2 \ln 2)} \qquad \dots(43)$$

and finally, because $p = \dfrac{t_d}{t_{dr}}$

$$b_{\frac{1}{2}} = 2\sqrt{(2 \ln 2)} \cdot \frac{t_{dr}}{\sqrt{n}} \qquad \dots(44)$$

Thus we obtain an equation by means of which the number of theoretical plates may be calculated from the chromatogram:

$$n = 8 \ln 2 \left(\frac{t_{dr}}{b_{\frac{1}{2}}}\right)^2 \qquad \dots(45)$$

The equation for the Gaussian curve of the peak is

$$P_u^n = \frac{1}{\sqrt{(2\pi n)}} \cdot \exp\left[-\frac{p^2 \cdot n}{2}\left(\frac{t - t_{dr}}{t_d}\right)^2\right] \qquad \dots(46)$$

for which equation 42 is substituted in equation 40, or after replacement of p

$$P_n^u = \frac{1}{\sqrt{(2\pi n)}} \cdot \exp\left[-\frac{n(t - t_{dr})^2}{2t_{dr}^2}\right] \qquad \dots(47)$$

All these equations are only valid for linear isotherms; in addition, spreading due to diffusion in the gas and liquid phases has not been taken into account.

Let us assume that it occurs only in the gas phase. Then the Gaussian peak in the middle will be increased on both sides by

$$\Delta x = \sqrt{(2D \cdot t_d)}$$

where D = diffusion coefficient of the substance in the carrier gas.

The distance Δx is covered in the time $\Delta x \cdot t_d/L$ at a gas flow rate of $u = Lt_d$ so that a spreading of

$$b_{\mathrm{diff.}} = \frac{2t_d}{L}\sqrt{(2Dt_d)}$$

results, where L = column length.

The total width at half height is thus

$$b_{\mathrm{total}} = b + b_{\mathrm{diff.}}$$
$$= \frac{2\sqrt{(2 \ln 2)}}{\sqrt{n}} t_{dr} + \frac{2t_d}{L}\sqrt{(2D \cdot t_d)}$$

In practice it is very useful to know the number of theoretical plates for the column being used, and if it is desired to calculate in advance the column performance necessary for a given separation then this knowledge becomes indispensable. We shall deal with the determination of the number of theoretical plates at a later stage, but here the reader's attention is drawn to an error which is continually being made in the use of the commonly used formulae and which can be avoided.

The number of theoretical plates is calculated by equation 45 using the values of $b_{\frac{1}{2}}$ and t_{dr} read off from the chromatogram.

From the derivation it is obvious, however, that one would have had to calculate the number of plates n from a value for $b_{\frac{1}{2}}$ which had undergone no spreading due to diffusion by the previously given value b_{diff}. But the values of $b_{\frac{1}{2}}$ available for measurement are practically already the value b_{total}.

Consequently a more correct formula must be calculated as follows:

The width of the air peak at half height $b_{d\frac{1}{2}}$ is measured. This air peak represents nothing but the minimum spreading due to diffusion of the substance in the gas phase. Now the width at half height of another peak which appears later and is easy to measure is taken for $b_{\frac{1}{2}}$. Further calculation gives:

$$n = 8 \ln 2 \left(\frac{t_{dr}}{b_{\frac{1}{2}} - b_{d\frac{1}{2}}} \right)^2 \qquad \text{....(48)}$$

Another corrected formula which takes into account diffusion in the gas phase in another manner was given by Purnell[27]:

$$n = 8 \ln 2 \frac{t_r \cdot t_{dr}}{(b_{\frac{1}{2}})^2} \qquad \text{....(49)*}$$

As we have already mentioned, a non-ideal relationship between the substance to be separated and the liquid phase has an effect on the shape of the peaks.

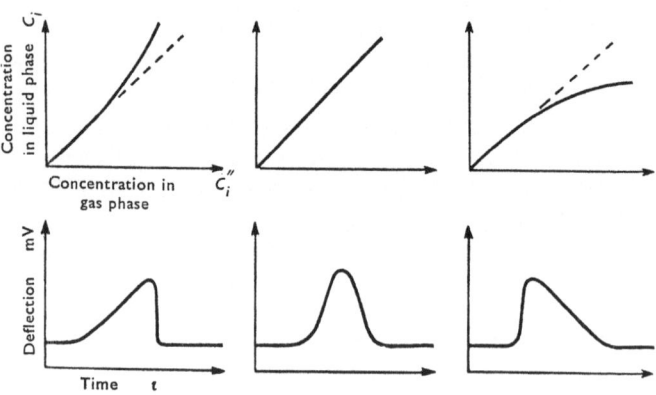

Figure 7. Relationship between the behaviour of substance i towards the liquid phase and the form of the peaks in the gas chromatogram.

* Equation 49 is more correct than equation 48.

Positive deviations from Raoult's law ($\gamma > 1$) gives peaks with a flat front and a steep back, taken in the direction of the time co-ordinate. Ideal behaviour gives almost symmetrical Gaussian forms. Negative deviations from Raoult's law ($\gamma < 1$) give peaks with a steep front and a flat back. (See *Figure 7*.)

As at the front of a zone moving through the column, especially in the middle, more molecules pass into the liquid phase (heat of condensation) than evaporate, this region is slightly warmer than the back, where more molecules pass into the gas phase (evaporate). Within the zone lies a temperature gradient which is the steeper the more substance is present. There are thus further influences overlying the process. The smaller the amount of substance to be separated, the closer gas chromatography comes to operating under ideal conditions.

Gas Chromathermography by the Circulation Technique

(Process according to Zhukhovitskii and Turkel'taub)

By definition this is a type of gas chromatography in which, instead of the entire column length being exposed to a temperature which increases with time, only a certain section of the column comes under the influence of an advancing temperature field. The difference from (linearly) programmed gas chromatography is as follows.

If the entire length of the column is subjected to a temperature programme on a straight line basis, then the logarithmic scale of the retention values of a homologous series will change into a straight line scale, i.e. the retention values always increase with the carbon number of the members of a homologous series; thus butane will be retained for 4 times and octane 8 times as long as methane.

But if the temperature field travels along the column the picture is altered. This is achieved by having a movable heater travel down the column in the direction of the column outlet. The dimensions of the heater are arranged so that between the front of the heater (e.g. 25°C) and the back (e.g. 250°C) a linearly increasing temperature field is set up in the section of the column covered. The individual substances in the column then move with the heater in quite definite and characteristic temperature areas. For example, in a given case methane might travel in the 25°C area, ethane in the 40°C area, propane in the 60°C area, *iso*-butane in the 75°C area and *n*-butane in the 80°C area. This so-called migration temperature depends on the type of substance, the type of column packing, the rate of travel of the heater and the flow rate of the carrier gas mixture. If we call the migration temperature T_w, then

$$T_{wi} = f(k, v, u, z)$$

T_{wi} Migration temperature of substance i

k Factor dependent upon the type of column packing

v Rate of travel of the heater

31

u Flow rate of the carrier gas mixture

z Carbon number of the substance in relation to the homologous series to which the substance belongs.

As the different sections of the column covered by the heater will always be at the same quite definite temperatures, the individual components will always be found in the appropriate temperature regions. Should the substances move forward into the colder regions of the column they will be (logarithmically) retarded, while if they hang back they will be strongly (and logarithmically) accelerated.

Finally the heater reaches an empty section of the column which leads into the detector. There the individual components are recorded in order of appearance. If the column is ring-shaped the heater can move continuously round it. Each time the heater passes the detector inlet a chromatogram is recorded. The essential point about this method is that in place of the customary carrier gas the entire mixture flows continuously through the column. No sample injector is needed. This elegant process is especially suitable for trace analysis, since the trace components are greatly enriched by the continual addition of substance. It is also suitable for the continuous industrial supervision of the composition of a mixture from methane to around heptane, but it can only be used for substances which can tolerate the final temperature of 200–250°C.

The form of the chromatogram corresponds to that normally obtained from gas solid or gas liquid chromatography by the elution technique, although this is actually a method using frontal analysis, and without the temperature cycles it would yield the so-called step chromatogram (Zhukhovitskii, Turkel'taub, Vagin[22–24]).

All the laws which have been derived and mentioned in this section are valid for both the basic chromatographic processes, for gas partition chromatography (GLC) as well as for gas adsorption chromatography (GSC).

The details which are necessary for practical application may be derived from the theory of the process; these details are given in the relevant sections.

References

1. BEYNON, J. H., CLOUGH, S., CROOKS, D. A. and LESTER, G. R., *Trans. Faraday Soc.*, 1958, **54**, 705.
2. GOLAY, M. J. E., *Gas Chromatography*, eds. Coates, Noebels, Fagerson, Academic Press Inc., New York, 1958, p. 1.
3. CREMER, E., and MÜLLER, R., *Mikrochemie*, 1951, **36/37**, 553.
4. JAMES, A. T. and MARTIN, A. J. P., *Biochem. J.*, 1952, **50**, 679.
5. JAMES, A. T., *Biochem. J.*, 1952, **52**, 242.
6. JAMES, A. T., MARTIN, A. J. P. and SMITH, G. H., *Biochem. J.*, 1952, **52**, 238.
7. JAMES, A. T. and MARTIN, A. J. P., *Analyst*, 1952, **77**, 915; *Brit. med. Bull.*, 1954, **10**, 170.
8. LITTLEWOOD, A. B., PHILLIPS, C. S. G. and PRICE, D. T., *J. Chem. Soc.*, 1955, 1480.

References

9. RAY, N. H., *J. Appl. Chem.*, 1954, **4**, 21.
10. PIEROTTI, G. J., DEAL, C. H., DERR, E. L. and PORTER, P. E., *J. Amer. Chem. Soc.*, 1956, **78**, 2989.
11. HERINGTON, E. F. G., *Vapour Phase Chromatography*, ed. by D. H. Desty, Butterworths, London, 1957, p. 5.
12. CONSDEN, R., GORDON, A. H. and MARTIN, A. J. P., *Biochem. J.*, 1944, **38**, 224.
13. RÖCK, H., *Ausgewählte moderne Trennverfahren zur Reinigung organischer Stoffe* (Selected modern separation processes for the purification of organic substances), Dr. Dietrich Steinkopff, Darmstadt, 1957, p. 74.
14. KEULEMANS, A. I. M., *Gas Chromatography*, Reinhold Publ. Corp., New York, 1957, p. 171.
15. MARTIN, A. J. P., and SYNGE, R. L. M., *Biochem. J.*, 1941 **35**, 91.
16. MAYER, S. W. and TOMPKINS, E. R., *J. Amer. Chem. Soc.*, 1947, **69**, 2866.
17. GLUECKAUF, E., *Trans. Faraday Soc.*, 1955, **51**, 34.
18. DEEMTER, J. J. VAN, ZUIDERWEG, F. J. and KLINKENBERG, A., *Chem. Engng Sci.*, 1956, **5**, 271.
19. KEULEMANS, A. I. M. and KWANTES, A., *Vapour Phase Chromatography*, ed. D. H. Desty, Butterworths, London, 1957, p. 15.
20. BRENNAN, D. and KEMBALL, C., *J. Inst. Petrol.*, 1958, **44**, 14.
21. PHILLIPS, C., *Gas Chromatography*, Butterworths, London, 1956.
22. ZHUKHOVITSKII, A. A., and TURKEL'TAUB, N. M., 'Use of the thermal factor in gas chromatography,' *Dokl. Akad. Nauk SSSR*, 1957, **116**, No. 6, 986.
23. VAGIN, E. V. and ZHUKHOVITSKII, A. A., 'The theory of the thermal adsorption separation of gaseous mixtures', *Dokl. Akad. Nauk SSSR*, 1954, **94**, No. 2, 273.
24. ZHUKHOVITSKII, A. A., TURKEL'TAUB, N. M., and SHVARTSMAN, V. P., 'Theory of chromathermography', *Zhur. Fiz. Khim.*, 1954, **28**, No. 11, 1901.
25. KLINKENBERG, A., and SJENITZER, F., *Chem. Engng Sci.*, 1956, **5**, 258.
26. KIESELBACH, R., *Analyt. Chem.*, 1960, **32**, 880.
27. PURNELL, J. H., *Nature, Lond.*, 1959, **184**, 2009.
28. HABGOOD, H. W. and HARRIS, W. E., *Analyt. Chem.*, 1960, **32**, 450.

Further Literature

JANAK, J., 'Chromatography in the gas-solid system,' *Ann. N.Y. Acad. Sci.*, 1959, **72**, 606.
GLUECKAUF, E., 'Movement of highly radioactive gases in absorption tubes (special problems in the GLC of radioactive gases),' *Ann. N.Y. Acad. Sci.*, 1959, **72**, 562.

A review of recent literature on theoretical problems will be found in *Gas Chromatography Abstracts*, 1958, 1959, 1960 and following issues (ed. C. E. H. Knapman, Butterworths, London) under Subject Index No. 2.

A particularly thorough and clear presentation of the theory of gas chromatography, including gas chromathermography which has been neglected in previous literature, is given by G. Schay's book *Theoretische Grundlagen der Gas-Chromatographie* (Theoretical Fundamentals of Gas Chromatography), Deutscher Verlag der Wissenschaften, Berlin, 1960.

2. APPARATUS

PRELIMINARY COMMENTS

ALL chromatographic methods require apparatus. But while a developed paper chromatogram is in existence for a given time, in a gas chromatogram everything is continually changing, and it is necessary at the first stage of separation to carry out the 'development' and the preliminary evaluation of the chromatogram at one and the same time. Special instruments are needed to record the separated substances, because, generally speaking, in gas chromatography these cannot be made visible, as in paper chromatography, or weighed after removal of the liquid phase, as in column chromatography. It is therefore important to record the chromatographic result at once. As a gas chromatogram results from the fact that the individual components, starting at a given time, move through the column at different speeds, it is necessary to measure not only the quantity of substance but also the time at which it leaves the column. It has been said that the classical methods of obtaining a chromatogram by making the components visible or by weighing them can generally not be used in the case of gas chromatography. The difference between gas chromatography and the other chromatographic processes with respect to the apparatus needed may be made clear from the example of the flame emission process, which will be described in more detail in the section on detectors.

If the carrier gas is combustible it may be burned at the end of the column in a small flame. Assume that a mixture of aromatics is to be separated. The individual aromatics move down the column at different speeds. As each aromatic has to travel the same distance, the fast moving components arrive in a short time and the slow moving components in a longer time at the column outlet and thus at the flame. The flame, which is burning only dimly (we shall assume that hydrogen is being used as carrier gas) burns up brightly each time an aromatic appears. We can thus 'see' the gas chromatogram. But in order to know the identities of the components it is necessary to record the times at which the individual aromatics were seen. From a paper chromatogram the migration rates of the individual components can be determined by two length measurements, but in the case of the gas chromatogram a time measurement is necessary.

The apparatus for gas chromatography generally consists of:

(a) a column. As, especially in the case of quantitative analysis, a reproducible or exactly known quantity of substance, which must be very small, has to be introduced without loss at the column inlet, a system for introducing the sample, a so-called sample injector, must be included in the apparatus, also

(b) a detector, which will measure and record the appearance of a substance together with the time at which it appears. The theory of GLC

34

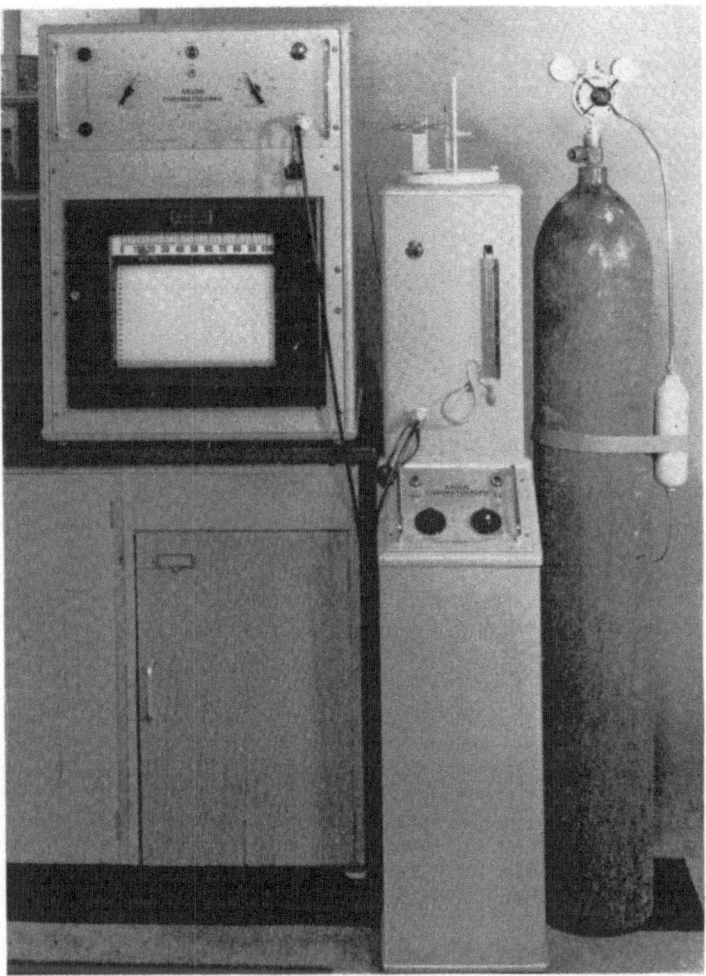

Figure 8. Example of an apparatus with gas cylinder, thermostat unit and recorder with associated electronic equipment (Pye Argon Gas Chromatograph); by kind permission of W. G. Pye Ltd., Cambridge.

requires that a constant stream of gas at a constantly defined temperature flows through the column. Therefore the apparatus also includes:

(c) a gas source,

(d) control and measuring instruments, which produce and control a constant stream of gas, and finally

(e) thermostats and temperature measuring and control instruments.

Under such conditions the separation and measurement processes proceed completely automatically. It is a recognized fact that gas chromatography is fundamentally suitable for partially or completely automatic operation. As in any case the exit of a substance from the column is measured in the form of physical values which may be converted systematically into electrical values, it is self-evident that a voltage recorder capable of automatically recording fluctuations of voltage with time will complete the gas chromatographic apparatus.

The paper supply to the recorder is a measure of the time taken. Without going into it any further, we can see that this causes a considerable increase in the costs of a gas chromatographic apparatus, but it leads to a process capable of automatic operation with all the associated advantages.

The minimum requirements for apparatus increase with the difficulty of the separation. The cheapest apparatus is that used only for the analysis of gases. The dearest is that used for the analysis of high-boiling substances.

Automation can be taken to the point where an integrator is connected either directly to the detector or indirectly through the recorder, and takes the quantitative values of the chromatogram and prints them together with the time of analysis or some other time-dependent factor.

An integrator which neither prints nor records the integral is an expensive and not particularly useful addition to the apparatus.

The many different methods used in gas chromatography make it almost impossible for the apparatus manufacturer to satisfy all the user's requirements.

There is no universal gas chromatograph. At present the economically most favourable solution appears to be the gas chromatograph built up from standardized units, which can be assembled in as many different ways as possible and thus may be suited to the job in hand. The present section aims to help the reader by giving him a more thorough knowledge of the apparatus used. Although there are in existence a number of first-rate instruments made by reputable firms, the versatility of gas chromatography can only be fully exploited if the user

(a) can recognize the type of instrument most suited to the job in hand

(b) has an exact knowledge of the functions and limitations of the parts of the apparatus—also from theoretical considerations—and

(c) can keep up with the rapid advances in technique by undertaking this or that constructional improvement himself (see the diagrams).

2.1. THE COLUMN

The following section gives information on the external form and the internal arrangement of the column. As the efficiency of gas chromatography depends to a decisive extent on the quality of the column, practical measures for the production of efficient columns are dealt with in some detail. The necessary properties of the solid support, the liquid phase, and the column impregnation are described. First, data are given on the quantitative description of column quality.

It is expedient to distinguish two types of column: packed columns and the so-called capillary columns (Golay[1]).

The present volume deals exclusively with the problems of packed columns. Capillary columns will be described in Volume II.

The Column Packing

In GLC this consists of a solid, porous, generally inorganic material which is impregnated with 1 to 40 wt. per cent of a liquid, the so-called liquid phase. (The liquid phase may be a solid at room temperature.)

Before we tackle the problems of the preparation of columns, there is the question of the evaluation of the column to be dealt with.

Evaluation of Packed Columns

There are four values that characterize the packed column both for GSC and GLC.

1. The column performance, which is measured as the number of theoretical plates per metre column length.
2. Selectivity with respect to a given separation, which is measured as the retention of the two components of the mixture most difficult to separate, and resolution.
3. The load capacity for the mixture to be separated, which is measured as the maximum quantity of mixture in mg (or γ) which when injected into the column will still allow a column performance of 90 per cent of the theoretical to be maintained.
4. The performance index of the column in poises. It is an empirical means of comparison which combines the number of theoretical plates, the pressure drop, and the time necessary for the analysis.

Values 1 and 3 only depend specifically upon the condition of the column when the measurement is carried out at a definite temperature (or over a definite temperature range) and with a definite carrier gas. Value 2 depends on the qualitative composition of the sample and the type of liquid phase. We find that the column cannot be evaluated without considering the separation to be accomplished, because it is just that relationship between column packing and substance to be analysed which is so close and specific. Value 4 gives a basis for comparing different columns, and in this the column performance, the dead volume and the flow resistance are combined into one value, the performance index. Further, the evaluation cannot be carried out without considering the operating conditions, because, for example, the number of theoretical plates is, among other things, a function of the carrier gas flow rate, and the selectivities depend upon the temperature.

The first three values give so valuable an assessment of the column that they should always be determined, so that a column is never used blindly. Only then can the optimum operating conditions be established and used.

Naturally there are other details of the column which influence the separation to be carried out. As has been mentioned in the section on theory, the shape of the peaks is dependent on other factors. Unsymmetrical forms may be obtained if there are interactions between the liquid phase and the substance, which can be explained as a deviation from Raoult's Law.

The Column Performance n' and the Separation Factor β

By column performance is meant the maximum number of theoretical plates in 1 metre of column length.

Although the theoretical plate concept originated from the classical theory of equilibrium analysis, this concept, together with that of *HETP*, may also be derived from statistical and non-equilibrium considerations of gas chromatography.

Calculation of Number of Theoretical Plates, Column Performance and Separation Factor

The number of theoretical plates is calculated from an analytical result which is obtained with the column. The substances and the operating conditions should be those which will normally be used in practice.

There are a number of formulae for the determination of the number of theoretical plates. Struppe[2] has shown that only 7 of these formulae are identical and has selected those suitable in practice[3-12]. In spite of the agreement of the 7 formulae it is expedient, in view of the method of evaluation to be used, to select the formula which gives the most exact values for the simplest measurement. For this reason the following formula was chosen.

$$n = 8 \ln 2 \left(\frac{t_{dr}}{b_{\frac{1}{2}}} \right)^2$$

$$= 5.54 \left(\frac{t_{dr}}{b_{\frac{1}{2}}} \right)^2$$

$$= 5.54 \cdot \beta^2$$

From this we obtain the first characteristic for the column packing, the column performance n':

$$n' = 5.54 \left(\frac{t_{dr}}{b_{\frac{1}{2}}} \right)^2 \frac{1}{L}$$

where L is the length of the column in metres.

The greater the retention time of the peak from which n is determined, the smaller will be the measurement errors of the determination of n' and n. For the calculation a peak with a relatively great height and symmetrical shape is chosen, which can safely be assumed not to have given rise to any overloading of the column.

The dependence of the number of theoretical plates on the length of the column is approximately linear (Scott[13]).

$$n = n' \cdot L = \text{const. } L$$

This relationship holds true more readily, the closer the ratio of the inlet and outlet pressures is to 1, i.e. $p_i/p_0 \rightarrow 1$.

In the above formula for the number of theoretical plates the actual measured value is the ratio of the retention time (uncorrected) to the average peak width. This value is called the separation factor.

$$\beta = t_{dr}/b_{\frac{1}{2}}$$

t_{dr} and $b_{\frac{1}{2}}$ are read off from the chromatogram in mm, cm, or sec. It is only the column length that has to be given in metres.

Selectivity with regard to Retention and Resolution

The selectivity of the column packing for a given pair of substances affects the retention. (For instance, the greater the selectivity for a meta/para-xylene separation, the greater will be the resolution.) This, together with the column performance, determines the degree of resolution (degree of separation) that can be achieved for a given pair of substances.

This is the second characteristic for the column. The value is given as a dimensionless number (always less than one) or, multiplied by 100, as a percentage.

The resolution is always related to a given pair of substances, generally the two substances most difficult to separate in a given mixture.

The definition and method of measuring and calculating the resolution can be seen from *Figure 9*.

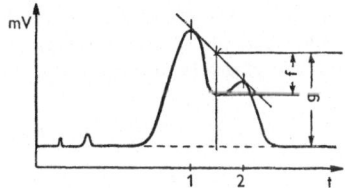

Figure 9. Determination of resolution
$\vartheta = f/g$

By a resolution of 0·5 or 50 per cent is understood one where two neighbouring peaks overlap at exactly half their height.

English and American authors often define resolution as the ratio of the time between the peak maxima 1 and 2 to the average peak width of peaks 1 and 2.

$$R_{21} = \frac{2(t_{dr2} - t_{dr1})}{\varDelta t_2 + \varDelta t_1}$$

t_{dr2} = retention time (uncorrected) for substance 2

$\varDelta t_2$ = width of peak 2 measured as the distance between the point of intersection of the tangents through the point of inflection with the base line.

For peaks which lie close to one another this definition may be simplified

$$R_{21} = \frac{t_{dr2} - t_{dr1}}{\varDelta t}$$

to where it is assumed that the peak width based on peaks 1 and 2 is equal.

This definition has a number of disadvantages and will not be used here. For the purposes of comparison in the reading of the literature the relationship between the two definitions is given: The following relationships are approximately valid for values between 0 and 0·999 . . .

$$\vartheta = 2R - 1 \text{ or } R = \frac{\vartheta + 1}{2}$$

Calculation of the Number of Theoretical Plates or Column Length Necessary for a Given Separation

The number of theoretical plates n and the separation factor β are each characteristic for the column. The column performance n' is characteristic for a given column packing. The selectivity, measured as the retention r for a given pair of substances, is also characteristic for the column packing. But while n', n and β are independent of the particular separation, both the selectivity and the resolution can only be given in terms of the separation to be carried out.

This may be given as: separation of meta/para cresol

isobutene/1-butene

methyl stearate/oleate

For such a task the length of the column and the number of theoretical plates can be calculated. Thus the degree of resolution can be determined to the desired extent.

For analytical separations with a very short time of analysis, ϑ for the most difficult pair to separate should be 50 per cent, for precise analysis it should be 90 per cent, and for preparative separations 99·9 to 99 per cent. The following relationships apply:

$$n = 2 \ln \frac{2}{1 - \vartheta} \left(\frac{\dfrac{k + 2}{k} \cdot r + 1}{r - 1} \right)^2$$

$$r = \frac{t_{r2}}{t_{r1}} = \frac{V_{o2}}{V_{o1}}$$

$$k = \frac{t_{r2}}{t_d} \qquad \qquad \text{....(1)}$$

(see comments under 1), and

$$n = 2 \ln \frac{2}{1 - \vartheta} \left(\frac{\alpha + 1}{\alpha - 1} \right)^2$$

$$\alpha = \frac{t_{dr2}}{t_{dr1}} \qquad\qquad ....(2)$$

(see comments under 2)

There are two methods of applying the relationships between retention, resolution and the appropriate number of theoretical plates.

1. By the use of tables. From tables of specific retention or relative retention time the retention of the pair of substances to be separated can be calculated. The liquid phase which is chosen will be the one that shows the greatest retention (= selectivity) for the pair of substances to be separated. The operating conditions chosen will then give a definite value of V_{r2} for the second component of the pair, and the column used will give a definite value for the column gas volume V_G. Thus all the required values are available, and all that is necessary is to insert the desired value for the resolution.

The following figures may be of interest:

for a ϑ of 50 per cent the pair of substances overlap by 12·5 per cent
for 80 per cent the pair of substances overlap by   3·2 per cent
for 90 per cent the pair of substances overlap by 0·9 per cent
for 99 per cent the overlap is only   0·0001 per cent

If the method is being used preparatively, the degree of overlapping is a measure of the purity of one of the two components.

To simplify the calculations, a few values for the factor $2 \ln \dfrac{2}{1-\vartheta}$ are given in the table.

$2 \ln 2/(1 - \vartheta)$	ϑ	ϑ per cent	Purity
1·38	0	0	0 per cent
1·63	0·1	10	
2·78	0·5	50	Contaminated with 12·5 per cent
6·00	0·9	90	Contaminated with 0·9 per cent
10·6	0·99	99	Contaminated with 1×10^{-4} per cent

2. From values which can be read off from the chromatogram and which refer to a pair of substances which have separated moderately well, we can calculate with the aid of equation 2 to what extent the column will have to be shortened or lengthened in order to achieve a desired degree of resolution.

Here α is the ratio of the uncorrected retention times.

Both α and r are always greater than 1.

By using the equations given, the number of theoretical plates may be calculated, as in the following example for a benzene/toluene mixture:

Liquid phase: succinate ester, 100°C
For 10 per cent resolution (just recognizable)   18 plates
50 per cent resolution (just usable for quantitative analysis)   32 plates
99 per cent resolution (suitable for preparative separation)   116 plates

With a packing that has a column performance of $n' = 1,000$, only a 4–12 cm long column will be needed for 50–99 per cent resolution. This is based on the assumption that the sample injector is as good as possible and is working under optimum conditions, and that the detector does not have any unsuitable dead volume.

As a basis for assessing the column performance of column packings, it may be noted that

for very good packings: $n' = 2,000–4,000$
for good packings: $n' = 800–1,200$
for moderate to poor packings: $n' < 600$

Between the *HETP* and the column performance n' there is the following simple relationship:

$$HETP = \frac{1,000}{n'} \text{ (mm)}$$

The highest number of theoretical plates that has ever been achieved with a packed column[13] is $30,000n$. The column in question had a length of 17 m, and thus the column performance of the packing used was 1,760.

It would be possible to prepare longer packed columns, but in view of the capabilities of the non-packed impregnated capillary columns there is no point in doing so. Of course it is practically possible to produce columns longer than 50 m. But if such a length is needed for preparative purposes it is more convenient to use the so-called circulation column proposed by Martin[14]. This consists of two columns of equal length, each connected to a flow-through detector, in which the carrier gas stream is always connected to the column in which the pair of substances is to be found. The carrier gas inlet and outlet are thus constantly changing.

α and r vary with temperature; according to Cvetanovic[15]

$$\log \alpha \text{ (and } r) = \frac{a}{T} + b \cdot \log T + c$$

a, b and c are constants.

Further details about the influence of temperature are given in Section 2.6.

As a rule, we may say that high-boiling liquids (b.p. above 200°C) are separated on short (2 m) columns. Low-boiling liquids and gases (b.p. below 200°C) need long columns.

Load Capacity

The third column characteristic is the load capacity. This can only be approximately calculated, depends on the substance and is best obtained by experiment.

Definition: the load capacity is the quantity in mg or ml of an individual substance which reduces the column performance to 90 per cent of the original value.

Performance Index

The fourth and last characteristic is especially useful for the quantitative comparison of different columns, and if necessary packed and capillary columns may be compared one with another.

The index I is obtained from analytical results. The necessary data are:

t_r the adjusted retention time of a substance appearing late and having symmetrical peaks, here measured in secs

t_d hold-up time in secs. Also in secs is the value for

$b_{\frac{1}{2}}$ the peak width at half height for the substance with a retention time of t_r

$p_i - p_o$ the pressure drop along the column in dynes/cm^2.

We may then say

$$I = \frac{(b_{\frac{1}{2}})^4}{t_r{}^3 \cdot (t_r - \frac{15}{16}t_d)} \cdot t_d \cdot (p_i - p_o) \quad \text{(poise)}$$

The performance index I is a measure of the quality of the type and arrangement of the column packing. It is an absolute column parameter, provided that the operating conditions are kept to their optimum value.

Time of Analysis and Resolution

In using gas chromatographic methods one should aim to carry out a given analysis with the required resolution in as short a time as possible.

A column with a large number of theoretical plates is not always the most suitable for this purpose. The relationship between the column parameters and the ratio of resolution to time of analysis (for optimum conditions this value must be large) is given, according to Loyd, Ayers and Karasek, by a calculation with the definition for ϑ used here where

ϑ = resolution

t_{dr2} = uncorrected retention time for substance 2 (min)

k_2 = t_{r2}/t_d

t_{r2} = retention time for substance 2

t_d = hold-up time

V_G = column gas volume, corrected to 0°C and for the pressure drop along the column

F_0 = gas flow in ml/min at 0°C

p_i, p_o = column inlet and outlet pressures

n = number of theoretical plates

$$\frac{\vartheta}{t_{dr2}} = \frac{3F_0\left(\frac{p_i}{p_o} - 1\right)}{2\left[\left(\frac{p_i}{p_o}\right)^2 + \frac{p_i}{p_o} + 1\right](1 + k_2) \cdot V_G} \left[\frac{k_2\sqrt{n}}{2(1 + k_2)} - 1\right] \quad ....(3)$$

The factors given in equation 3 have a restricting effect on one another. The type and film thickness of the liquid phase, however, influence both the column performance and the retention times. The factor $k_2/(1 + k_2)^2$ contained in equation 3 is also influenced by the liquid phase; it has a maximum value for $k_2 = 1$.

The van Deemter equation as modified by Jones and Kieselbach (see p. 21) specifies that the column performance is independent of the type of carrier gas, provided that the influence of diffusion in the liquid phase is kept small relative to the other values, which may be achieved by reducing the film thickness. All this enables us to predict that for rapid analyses a carrier gas of low viscosity and a column packing of low impregnation should be chosen. Further, Purnell and Quinn[38] have discovered that the choice of liquid phase and separation temperature has a decisive influence on the obtaining of minimum separation times. The liquid phase should be so constituted that the two substances to be separated have partition co-efficients of over 50 and k_2 does not exceed the value 2·0.

If it is not possible to find liquid phases and conditions which give sufficiently large partition coefficients for the substances, then the column must be shortened.

For almost complete resolution ($\vartheta \sim 0.99$) and for values of $k_2 = 2$ the time needed for the separation of a pair of substances with retention r is:

$$t = 14\cdot6 \,.\, 10^3 \,.\, \frac{t_d}{n^1 \,.\, L} \left(\frac{r}{r-1}\right)^2 \text{(sec)}$$

n' = column performance
L = length of column in m
t_d = hold-up time in min
r = retention

Generally, where $k_2 = 2$ and the resolution is almost complete, this expression may be put:

$$t = 243 \left(\frac{r}{r-1}\right)^2 (C_g + C_l)$$

The value for $C_g + C_1$ may be obtained from the relationship expressing the variation of *HETP* with gas flow rate u_0 (measured at the column outlet in cm/sec). This expression is evaluated graphically, using a series of experiments to provide the necessary values.

The factors of importance for rapid analysis may be summarized as follows:

1. The concentration of liquid phase in the column packing should be no higher than 10 wt. per cent.
2. The liquid phase and operating temperature should be related to the separation in hand in such a way that the pair of substances most difficult to separate have the maximum possible retentions and the components have partition coefficients greater than 50.
3. The carrier gas should have as low a viscosity (hydrogen) and as high a flow rate as possible.

4. The column length should be so adjusted that the most difficult separation can still be carried out with a resolution of over 50 per cent.
For obtaining optimum separation conditions see also Giddings, J. C., *Analyt. Chem.*, 1960, **32**, 1707.

Column Material

Suitable materials include refined steel for rigid straight or U-shaped columns, copper for columns whose external shape must be altered after they have been packed (e.g. coiled), hard glass for columns which must not give rise to catalytic wall effects, and plastics for columns for the analysis of highly aggressive substances such as $(FH)_2$, $POCl$, etc. The wall thickness of plastic columns must not be less than 1 mm. Copper, which is otherwise very suitable as a column material, has an active surface which interferes with many separations. This factor may be eliminated by silver or gold plating. For silver plating the column is first rinsed well with dilute nitric acid and then a solution of silver cyanide complex is passed through for about 10 min at room temperature. A suitable complex solution may be prepared by the hot saturation of a 25 per cent aqueous KCN solution with AgCl. The conditions for gold plating are somewhat more difficult.

Column Diameter

For analytical purposes, in which the possibility of the identification of unknown substances must still be allowed for, a column diameter of 6 mm is sufficient. For purely analytical separations in which it is not necessary to obtain the substance, columns of 3 mm diameter are satisfactory.

The load capacity grows quadratically with the column diameter (for columns with equal loading with liquid phase, see p. 55).

$$B = k' . d^2$$

where B is the load capacity in mg, d is the internal column diameter, and k' is a constant.

For preparative purposes columns with internal diameters of between 10 and 25 mm are therefore used.

Bayer[31] was able to show that even columns with diameters of 100 mm could be used for preparative purposes. By means of directional members within the column packing it was possible, in spite of the very large diameter, to maintain a column performance of $n' = 500$.

The Solid Support

Basically any solid, inert, granular material can be used as solid support. Suitable substances are: kieselguhr, burnt and unburnt clays, zeolites, metals, glass, sand, salts, powdered plastics, and for special purposes active adsorbents.

The only considerable difference in the individual materials is in the column performance n'.

This is based on the assumption that the liquid phase film is spread in an optimum fashion over the support.

The values of n'' for various materials are approximately as given in the table on page 47.

Certain conclusions may be drawn from the above. The best solid support is granular, specially selected kieselguhr (Chromosorb). But unburnt clays are also very effective; on the other hand burnt clays are only about one fifth as good. The main reason for these facts may be found in the surface structure. Home-made solid support with optimum pores and minimum pressure drop is also very good.

The solid support must have a minimum surface area of *ca.* 1 m²/g. Materials suitable are not those with long narrow fissures, but rather those with shallow pores, as round as possible, which can be coated with a very thin film of liquid phase. This very structure is found in certain types of kieselguhr. The shells of minute marine animals represent the optimum form. A laminar structure perforated with many small holes is to be preferred to the shallow and deep-pored silica gel, although this has a hundred times the surface.

Since in GLC the support should generally not play an active part in the separation process, an inert material (SiO_2) is to be preferred to an active material ($SiO_2 . nH_2O$). The same relationships are found with clay. Burnt clay has lost much of its surface in the sintering process; it is certainly less active than unburnt clay, but only gives low n values. Unburnt clay has a greater surface area, but its chromatographic adsorption activity is much too great.

A spherical support which consists of a porous material with absolutely no adsorption activity and which has pores of 0·001 mm or rather less may be described as especially good, indeed ideal. This porous material is formed into spherical particles and made mechanically hard. This is achieved by mixing the finely powdered raw material with small amounts of inorganic binders. The mixture is first formed into spherical particles on obliquely turning plates (as in the production of pills) and then burned. Diameter of spheres: no greater than 0·275–0·300 mm[33] (Leibnitz[40]).

Ettre[39] used Nelsen and Eggertsen's method (*Analyt. Chem.*, 1958, **30,** 1387) for a comparative determination of the surface areas of various supports. He found for

Firebrick	4·14 m²/g
Chromosorb P	2·86 m²/g
Chromosorb W	1·41 m²/g
Celite	1·14 m²/g (best result for n' and ϑ)
Teflon powder	0·23 m²/g (worst result for n')

Ettre found that the specific surface area did not affect the column performance so long as it did not fall below a certain limiting value (1 m²/g). In spite of the similar particle size firebrick causes a greater carrier gas pressure drop than Celite or Chromosorb, so that the particle form must also

Approximate values of n′ for solid support materials

Material	n′	Comments
1. Kieselguhr, coarsely granular, Chromosorb brand, most suitable type	3,000–5,000	Surface area: 3 m²/g, gas flow resistance low. Loadable with 30 wt. per cent liquid phase.
2. Kieselguhr, suitable type, purified, without clay	1,800–2,000	Gas flow resistance very high. Loadable with 30 wt. per cent liquid phase.
3. Johns Manville firebrick C22	1,600–2,300	3 m²/g surface area, otherwise as for 1.
4. Sterchamol No. 22 (a kieselguhr-clay mixture) of type suitable for gas chromatography	1,000–1,500	Gas flow resistance low. Loadable with 25 wt. per cent liquid phase.
5. Unburnt clay of type MEKA (mechanically prepared *ka*olin)	700–1,300 1,100–1,300	Loadable with 20 wt. per cent liquid phase. Loadable with 20 wt. per cent liquid phase.
6. Silica gel, wide pores Silica gel, narrow pores	200–400 100–300	Loadable with 60 wt. per cent liquid phase. Surface area 340 m²/g.
7. Burnt clay (brick, firebrick, chamotte)	200	Loadable with 8–10 wt. per cent liquid phase.
8. Steel in the form of Braunschweiger spirals 1·5 × 1·5 mm	40–50	Loadable with 0·5–1 per cent liquid phase. Gas flow resistance, especially in negative pressure region, very low. Δp = 2–10 mm Hg/5m.
9. Steel in the form of Braunschweiger spirals, 2·5 mm × 2·5 mm	6–10	
10. Table salt	20–80	Loadable with 1–3 per cent liquid phase for high temperatures at negative pressure.
11. Coarsely granular powder from teflon, Hostaflon	50–200	Loadable with 3–16 per cent liquid phase; for highly aggressive media.
12. Glass beads	50–200 according to diameter	Loadable with 1–3 per cent liquid phase.
13. Aluminium powder	50–100	Loadable with 3–5 per cent liquid phase.
14. Washing powder (with admixture of attenuated inorganic sulphonates, dried and washed with petroleum ether)	up to 1,000	Loadable with 10 per cent liquid phase, completely inactive.
15. Active adsorbent	up to 3,000	Loadable with 0·5–60 per cent liquid phase.

play a part. For strongly polar substances even the most effective supports were, because of a strong tailing, inferior to teflon, which is specially suited to this task.

Removal of Residual Adsorption Activity

While active supports are used in GSC and indeed, many gas mixtures are more readily separated when working with materials which have been given added activity (e.g. activated alumina, active magnesium silicate, etc.; to this group also belong supports based on active adsorbents which are only impregnated with small quantities of liquid phase[27, 28]), any residual adsorption activity interferes with GLC. It is especially troublesome in the following cases:

    (a) where the wetting with liquid phase is slight (1–10 per cent);

    (b) when weakly polar or non-polar liquid phases are used;

    (c) in the analysis of strongly polar substances.

In the case of kieselguhr and clay, the residual activity may be removed by exhaustive treatment with strong mineral acids, either singly or as mixtures, after rigorous drying with dichlorodimethylsilane[33]. Such supports give packings that give symmetrical peaks even with strongly polar substances, and in comparison with untreated supports the time required is reduced by a factor of 3. For the separation of strongly basic substances, detergents (inorganic fillers, aryl alkyl sulphonates, etc.) treated with KOH have been shown to be good solid supports[36].

Particle Diameter

Numerous investigations have been undertaken to discover the optimum granule size. Recently Scott[13] and Johns[16], *inter alia*, have tackled the problem. Earlier studies of Dimbat *et al.*[17], and Keulemans and Kwantes[18] have shown that the finely granular kieselguhr (type Celite 545) with particle diameter of 0·02–0·04 mm originally used by James and Martin could be replaced with advantage by the coarse-grained kieselguhr type Sterchamol No. 22 and similar insulating materials.

In the choice of the correct particle size two factors play a part: the effective surface area and the flow resistance. The coarser the particles, the smaller is the gas flow resistance, but at the same time the smaller is the effective surface area. The two values are opposed, and therefore there exists an optimum value.

For packed columns of high efficiency Scott recommends the use of granule sizes of 0·10–0·12 mm. Certainly the retention times of substances on such columns are very high, and for columns of length 15 m the inlet pressure is 12–14 atm. When a packing with a particle size of 0·3–0·4 mm is used with a similar column length and under otherwise comparable operating conditions, the inlet pressure, related to the optimum gas flow, is 2 atm.

For normal routine analysis it is important that the time needed for effective separation should be very low. In such cases particle sizes of 0·20–0·30 mm are required.

Particle Diameter

Figure 10 shows how the retention time of a component increases with decreasing granule size of the support, the gas flow being kept constant.

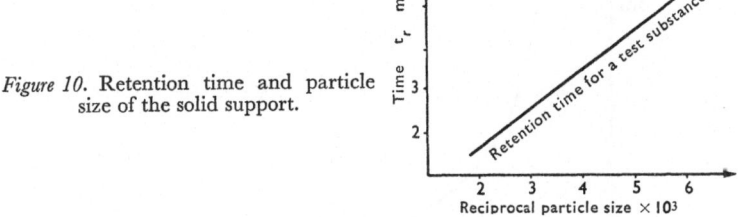

Figure 10. Retention time and particle size of the solid support.

Figure 11 shows how the pressure drop along the column (and thus the gas flow resistance for constant gas flow) increases with decreasing granule size, but the gradient of the curve for column performance continually falls away until no further improvement can be achieved. *Figures 10* and *11* are taken from the paper by Johns[16].

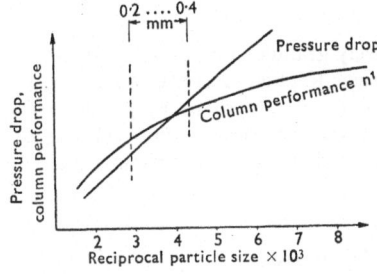

Figure 11. Pressure drop, column performance, and particle size of solid support.

It is clear from *Figure 10* why special care must be taken when selecting the particle size for very close-packed regions. A material which contains noteworthy amounts of particles of size 0·05–0·2 mm and also of size 0·4–0·6 mm will give a very bad column performance, because the retention time is determined by the particle size and the fine and coarse particles will cause considerable spreading of the peaks.

Care should therefore be taken, especially during impregnation, that the fine grain fractions produced by abrasion are removed. As in the literature the particle sizes are often quoted in B.S. mesh values, *Figure 12* is provided to enable them to be converted to metric values (mm).

A sieve analysis of a solid support of 0·2–0·4 mm particle size should show that at least 90 per cent of the granules lie within 0·1 mm of the range of values, and that deviations of more than ± 0·15 mm from the mean do not occur. For supports of granule size 0·10 mm the deviation from the average value should be no greater than ± 0·01 mm. Baker *et al.*[32] state that the optimum value is found in the region 0·257–0·300 mm.

The precautions to be taken when filling the column will be given later.

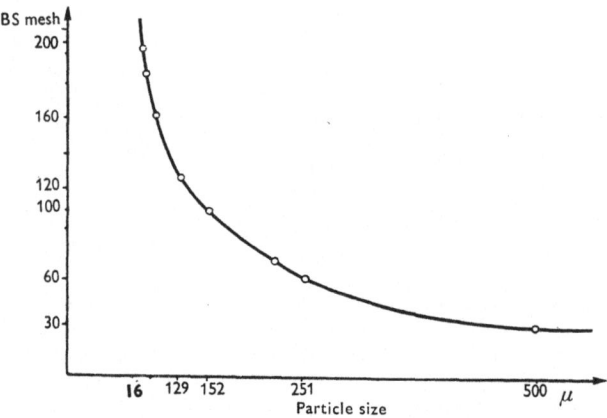

Figure 12. B.S. mesh values and metric measurements (mm⁻³).

If the method of crushing the coarsely granular material is correct (jaw crusher) then a yield of the desired sieve fraction of at least 55–60 wt. per cent should be possible. For vacuum gas chromatography Sorensen and Soltoft[19] have suggested the use of 1·5 × 1·5 mm spirals of fine steel wire and Cropper and Heywood[20] have suggested that table salt crystals should be used as solid supports.

The Liquid Phase

The liquid phase must fulfil five conditions.

1. For the temperature at which it is to be used, its vapour pressure must not exceed 0·5 mm. (But in special cases liquid phases can be used with vapour pressures of over 10 mm at the operating temperature.)

2. At this temperature the viscosity of the liquid phase should be as low as possible.

3. It must not undergo chemical (irreversible) reactions with the solid support, the carrier gas or the substances to be separated.

4. Thermal ageing of the liquid phase should lead to a stable state in as short a time as possible (low-boiling components should distil off, etc.)

5. Besides having optimum separating properties (selectivity) the purely physical properties of the liquid phase should be such that a completely uniform, firmly adhering film of stable structure can be produced and maintained.

The Liquid Phase

For good operation of the apparatus (particularly in the case of an automatic apparatus) adherence to conditions 1 and 4 is important; to obtain true results in quantitative analysis condition 3 is important; column performance and resolution are dependent upon conditions 2 and 5.

The choice of the most suitable liquid phase is the decisive problem in the application of gas chromatography.

On the one hand it is simpler to separate materials by using the right liquid phase than by the use of extreme conditions, on the other it is simpler to perform a given identification by using the right liquid phase than by using special measures, such as isolation of the separated components and identification with further chemical and physical methods. A special advantage of gas chromatography is that the analyst can use the specific action of different types of liquid phase either simultaneously or else one after the other. He is thus using time as a specific analytical factor. This factor is available to him by use of the column length and temperature in a predeterminable manner. In other words, at a given time after the start of the analysis a different column is introduced into the separation, and after a further time taken out again.

Figure 13 shows the layout for a two stage gas chromatograph, whose columns can be switched to parallel or series positions without causing any disturbance (Simmons and Snyder[21]).

Figure 13. Gas path arrangement for a two-stage gas chromatograph (according to Simmons and Snyder[21]).

S_1 = column 1.
S_2 = column 2.
V_{1-4} = on-off valves 1 — 4.
M, N = needle valves.
$A_{1,2}$ = gas outlets to detectors, 1, 2
P = sample injector.

Operating conditions:
pressure drop along column 1 =
  pressure drop across needle valve M
pressure drop along column 2 =
  pressure drop across needle valve N

If the columns are to be run in series, V_2 and V_4 are shut, V_1 and V_3 are opened.

If the columns are to be run in parallel, V_1 and V_3 are shut, V_2 and V_4 are opened.

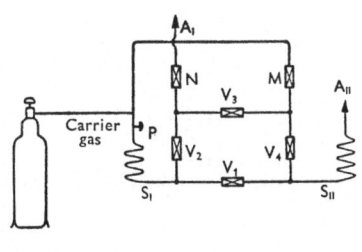

The advantage of the two or multistage method of operation is based on the fact that the substances which have been separated cannot possibly be remixed (provided that the gas paths are separate), even though they have to go through other columns for further separation from impurities of different structure. If the columns are arranged in series there is often the danger that components which have separated will catch up with one another again (in other words, remix—here we are talking in terms of rates of travel).

Figure 14 shows the analysis of C_5–C_8 hydrocarbons with a two stage instrument. As a switch valve a gas diaphragm valve may readily be used (see under sample injectors).

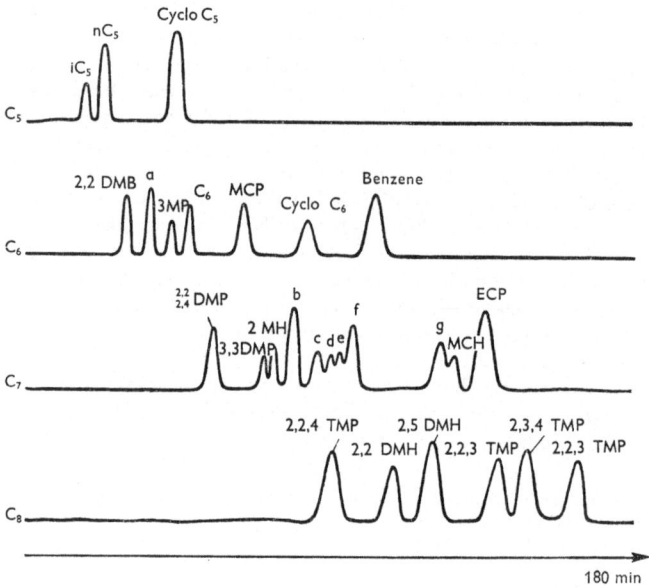

Figure 14. Analysis of 26 hydrocarbons on a two stage instrument according to Simmons and Snyder[21].

a = 2,3 DMB and 2 MP.
b = 2,3 DMP and 3 MH.
c = 1,1 DM cyclo P and 3 EP.
d = 1,3 DM cyclo P
e = nC_7
f = 1,2 DM cyclo P.
g = 1,1 DM cyclo P and M cyclo H
M = methyl; E = ethyl; D = di; T = tri; B = butane; P = pentane; H = hexane.

In the preparation of columns with different liquid phases it is important to realize that it makes no difference whether the different packings are situated in definite regions of the column, or whether, as a result of previous inner mixing, the different liquid phases are in the form of a single complex column packing (McFadden[22]). Naturally, the different liquids must not react with one another.

For the experimental determination of the optimum composition of complex columns it is more advantageous to proceed from the column packings of the initial fluids and to obtain the values from the most favourable length ratios of the individual columns.

For the separation of mixtures of substances belonging to several different chemical groups, it is very often best to use combinations of columns containing liquids of varying polarities (at least one non-polar and one polar liquid, in series or in parallel) (Keulemans *et al.*[23], Fredericks *et al.*[24]).

The rule given earlier that the correct choice of liquid phase is more likely to achieve success than the use of very long columns or complicated procedures will now be confirmed by an example, which at the same time provides an answer to the question often asked, namely whether it is always possible to predict the correct liquid phase for a given task.

The analysis of liquid hydrocarbons is, among other things, important in the investigation of the reaction products of isomerization, alkylation, polymerization, dehydrogenative and hydrogenative cracking, etc.

Figure 14 shows that, to take an example, the separation of 2, 3-dimethylbutane and 2-methylpentane (a) has not been successful, even though a two stage gas chromatograph was used. The column consisted of squalane (2·5 per cent) on coarse carbon black (Pelletex). The analysis of the C_6 hydrocarbons alone lasted two hours, the column length was 18 m, the operating temperature was 100°C. But if columns of 30 per cent heterocyclic amines (quinoline-quinine 7: 1 or *iso*quinoline) on kieselguhr are used the analysis of the isomeric hexanes can be accomplished in 25 min at 20°C, and 2, 3-dimethylbutane is completely separated from 2-methylpentane within 15 minutes.

Such liquid phases are also most suitable for the analysis of mixtures of naphthenes and alkanes (Zlatkis[25]).

In the choice of the most suitable liquid phase, i.e. the one with the highest selectivity for the given task, note must be taken of the forces which play a part in the mutual interaction between the components of the mixture and the liquid phase.

These forces may be divided into four groups:

(a) *London forces*. These are intermolecular forces between two non-polar substances. They are due to synchronization of the electronic oscillations of the two substances.

(b) *Keesom forces*. These are orientation forces which arise from the interaction of permanent dipoles. The length and size of the dipoles determine the value of the intermolecular forces; among these are the hydrogen bridges.

(c) *Debye forces*. These are due to induced dipoles. A permanent dipole induces dipole forces in neighbouring molecules of different composition. The size of such forces is only slight.

(d) *Chemical bonding forces*. These include all types of complex formation where the complex is unstable. In this the temperature at which the interaction between substance and liquid phase occurs plays an important role— the more stable complexes become increasingly unstable with rise in temperature. The strongest forces are those given under (a), (b) and (d).

The greater is the sum of the interacting forces, the greater the retention times will be. In this way the chemist has control over the type of selectivity. Non-polar substances on a non-polar liquid phase are separated in order of boiling point, as the intermolecular forces within the liquid phase and within the components are of the same order and kind. Thus nothing is altered when the liquid phase comes into contact with the components, and the boiling points of the substances determine their gas chromatographic behaviour.

Polar substances travel through a non-polar liquid phase more rapidly than do non-polar substances of the same boiling point, since the relatively strong dipole-dipole forces between the molecules of the polar substance can have no effect on the non-polar liquid phase. With increasing polarity of the liquid phase the polar substances are retained more and more strongly until finally the order of appearance of compounds of greater and lesser polarity is reversed.

Investigations of this type not only assist the practical application of gas chromatographic methods, but also in a very short time yield comprehensive numerical data, which among other things lead to the discovery of many new complex formers and highly selective liquids. Moreover, exact data may be obtained on the interacting forces in relation to the type and position of the functional groups.

Thus, James and Martin[29] found that when benzyldiphenyl is used as liquid phase the retention values of compounds of type $C_5H_{11}X$ increase with the different functional groups as follows:

$$X = H < Cl < Br < I < NH_2 < NO_2 < OH < COCH_3 < CN$$

On paraffin oil or a similar non-polar liquid phase the following series is found:

$$X = H < OCH_3 < Br, Cl < I < NO_2 < COCH_3 < NH_2 < CNH < OH$$

Such knowledge is especially useful in qualitative identifications.

Quantitative values, especially those concerning the strength of metal salt complexes, are also easy to obtain and use for difficult separations. (See also: Calculation of the retention indices of aliphatic, alicyclic and aromatic compounds, by Wehrli and Kovats[37], and in Volume III, Tables.)

The strengths of the interacting forces from the point of view of functional groups, active hydrogen atoms and strong polarities can be divided into five groups:

1. The molecules form three-dimensional hydrogen bridges (water, poly-alcohols, amino alcohols, oxyacids, polyphenols, di- and polycarboxylic acids).

2. The molecules possess both active H atoms and electronegative functions with free electron pairs (O, N, F). (Alcohols, fatty acids, phenols, primary and secondary amines, oximes, nitro compounds, nitriles with α-H atoms.)

3. The molecules possess electronegative functions but no active H atoms (ethers, ketones, aldehydes, esters, tertiary amines, nitro compounds, nitriles without α-H atoms).

4. The molecules possess H atoms, but have only a weak dipole character ($CHCl_3$, CH_2Cl_2, CH_3CHCl_2, etc., aromatics, olefins).

5. The molecules do not possess any functional groups and have no dipole character (hydrocarbons, CCl_4, etc.).

However, all the factors which have been mentioned in this section are still not enough to enable the most suitable liquid phase for a given separation to be chosen with absolute certainty. There are still gaps in our knowledge, which the chemist must bridge by instinct and by his knowledge of the materials. Numerous data on this are found in Volume III, Tables for Gas Chromatography.

Quantity of Liquid Phase

Two factors work in opposition when the quantity of liquid phase (generally given as grams on 100 g support) is to be varied; with increasing quantity of liquid the column performance generally falls, but the load capacity of the column rises.

Purely analytical tasks, in which the substance does not have to be frozen out of the gas stream or, indeed, removed in any way, should therefore be performed with packings of low impregnation, i.e. 5 g liquid phase on 100 g support. The load capacity is then about 0·1 mg per component for a cross-sectional area of the column of 10 mm².

For purely preparative purposes it is best to have packings with 20 g liquid on 100 g support. The load capacity is about 5 mg per component and 10 mm² cross-sectional area. The maximum possible quantity of fluid on a support depends on the nature of the latter. Silica gels can take up to 60 g/100 g, kieselguhr up to 40 g/100 g, burnt clays up to 12 g/100 g, glass beads up to 5 g/100 g, steel spirals up to 3 g/100 g, silicone rubber powder up to 5 g/100 g, etc.

From this set of figures alone it can be seen that the method of quoting results as g/100 g support is incomplete. We should really speak of g/m²

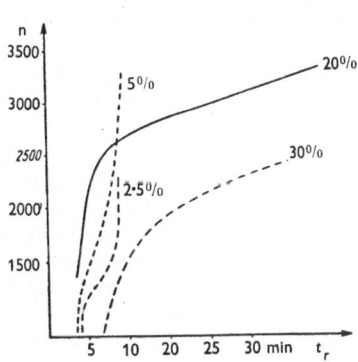

Figure 15. Retention time and number of theoretical plates vary with the quantity of liquid phase used, as a result of critical load capacity limits and other factors (from Cheshire and Scott[26]).

surface, since the specific weights of the individual supports are different. But the necessary information is often not available, and hence the inexact method of reporting is retained. Normally, however, only the coarsely granular kieselguhr should be used as support, then the weight data may be compared with the liquid phase. The impregnated packing must, of course, be absolutely dry; sticking means that the support is too heavily loaded with liquid phase, which leads to a sharp decrease in column performance and resolving power.

We should not overlook the fact that for packed columns of high efficiency, containing packings with low liquid phase loading, detectors of very high sensitivity are also required. Even columns with a load capacity of 0·1 mg/component and 10 mm² cross-section area require detectors with a sensitivity of $E \sim$ 40,000 (see p. 120).

The relationship between retention time, column performance and concentration of liquid phase on the support is shown in *Figure 15*. We can see that columns with 5 or 2·5 per cent liquid phase on the support show increasing column performance with increasing retention time, but this tendency is not so greatly developed with packings with 20 per cent liquid phase (Cheshire and Scott[26]).

Impregnation of the Support

The aim is to cover the entire surface of the support with a completely uniform and chemically and physically homogeneous film. It should be possible to carry out the impregnation quickly and reproducibly. Struppe[2] recommends the following method.

A sintered filter G_4 with a diameter of 15 cm is used as the impregnation vessel. Gas (compressed air, nitrogen) is blown through the stem of the funnel at a rate of *ca.* 100 l./hr. The funnel is heated evenly on all sides, either electrically or by placing in a water bath. The bottom of the funnel is also heated (electrically). While the gas is flowing, 50 to 250 g of the dry, sifted support are tipped into the funnel. The calculated quantity of liquid phase, diluted with petroleum ether, ether, ether-acetone, benzene, or carbon tetrachloride, is then added and gently stirred. The choice of a correct solvent is decisive. It must dissolve the liquid phase completely, so that no deposition is possible during the concentration process. The quantity of diluent used is calculated in such a way that at the start of the impregnation process the liquid phase completely covers the solid phase. The temperature of the funnel is controlled in such a way that the whole process is complete within 10 minutes, i.e. a dry powder remains behind. For most of the time the mass is no longer stirred. After 10 minutes the temperature of the funnel is raised by 20°C and the mass is dried for another hour at the same gas flow rate; the column packing is finally dried for 2–4 hours in a desiccator. The wash gas must of course be pure and above all dry (P_2O_5!) (see *Figure 16*). This method gives a substantial yield of good quality column packings, and is readily reproducible. As the material is subjected to hardly any mechanical stress, supports liable to abrasion may be treated. The impregnation only takes a short time.

If insoluble substances, such as teflon powder, or substances which are only soluble in water, such as sugar, are to be deposited on the support, then the 'liquid' phase is finely pulverized, intimately mixed with the granulated carrier, and sintered. A preliminary poisoning of the adsorption centres of the support is, however, necessary in this method (see below).

Another method of producing an even coating of liquid phase (especially polymeric liquid phases with not very good solubilities) is given by Horning et al[33]. This method makes use of the adsorption forces of the support. A

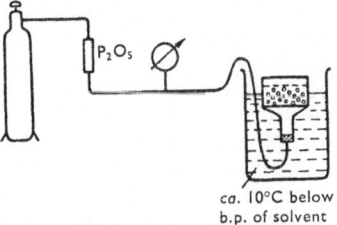

Figure 16. Plan of an apparatus for the reproducible impregnation of solid support (from Struppe, Kaiser and Leibnitz, *Gas-Chromatographie 1958*, ed H. P. Angele, Akademie-Verlag, Berlin, 1959).

ca. 10°C below b.p. of solvent

solution of the liquid phase (e.g. succinate polyester in acetone) of relatively low concentration is produced. 7·5 g of support de-activated with dichloro-dimethylsilane is introduced, with constant stirring, into 100 g of this solution. The suspension is kept in motion for a further 15 minutes. The adsorption equilibrium is then set up, and the solid material may then be filtered, gently pressed, or centrifuged out. It is carefully dried.

A 20 per cent impregnation may be attained with an 8 per cent solution.

A 9 per cent (or 4 per cent) impregnation may be obtained with a 4 per cent (or 2 per cent) solution.

In the production of the film of liquid, the polarity of the solvent plays an important role. The more weakly polar the solvent, the smaller is the danger that the liquid phase will be deposited on the support in droplet form (Szekeley et al.[34]).

The interface and surface tension values must be correctly influenced.

Most of the liquid phase is in the fine pores of the support, and only a small remainder covers the surface. Values for the thickness of the liquid film must thus be understood to be average values.

Testing the Column Packing

Peak Form

Highly unsymmetrical peaks or peaks that show a high degree of tailing have a strong negative influence on both the column performance and the resolving power of the column. Tailing is shown with particular readiness by strongly polar substances (alcohols, acids, amines, etc.).

This phenomenon results from free adsorption-active surface. But even when the entire substance is moistened, tailing is, in many cases, still found. This is the case when the natural strongly polar impurities in the liquid phase are not sufficient to 'poison' all the active area of the adsorptive surface. In order to complete this process the support must be treated alternately with mineral acid and alkali, washed with a large amount of pure water, and then finally 2–3 wt. per cent of a 'surface poison' (e.g. alkali soaps of C_6–C_{10} fatty acids, polyethylene glycols, long-chain amines of low volatility, esters of low volatility such as oleates, stearic acid, etc.) is added to the liquid phase; silver plating is also sometimes effective (Scott).

The direct de-activation of the surface with chlorosilanes has already been mentioned; it can follow the acid-alkali treatment directly after careful drying.

For the analysis of hydrocarbons on adsorbents such as active carbon, all that is needed to eliminate tailing is a simple impregnation with, for example, 1·5 per cent squalane.

Generally speaking, the pH value of the support does not play any part. However, it is of importance when acidic or basic substances are to be separated. Thus in the analysis of the lower amines it is expedient to moisten the support with a little sodium methoxide (2 per cent of the liquid phase) and for analysis of acidic substances the support should be impregnated with stearic acid or other acid compounds.

Control of the Quantity of Liquid Phase

Normally the quantity of liquid phase consumed is controlled by an ash determination on the support (blank value) and an ash determination on the impregnated packing.

Temperature Load Capacity

The maximum temperature for a column is that at which 0·5 per cent of the liquid phase is lost through evaporation within 10 hours at a gas flow rate of 15 ml/min. (Tuey, 'Material for gas chromatography,' *May and Baker Lab. Bull.* (Dagenham), 1958, 8).

This value corresponds to a concentration in the carrier gas of 1 mg/l. If an ionization detector is being used, then the concentration of evaporated liquid phase in the carrier gas must not exceed 0·01 mg/l. (Gerrard[35]).

Even the slightest traces of oxygen in the carrier gas endanger almost any liquid phase at temperatures above 100°C.

Filling the Column

Correct filling is especially important. Columns that are to be formed into a helix should generally be filled after the forming operation.

The material must be packed so tightly that, in spite of vibration in the packing due to carrier gas flow, no fissures or channels are formed, even when the column is horizontal. It is true that the column performance increases with increasing packing density, but the flow resistance of the column also increases greatly. The column must therefore be packed in a reproducible manner. For this purpose the lower end of a straight column, or the bend of a U-shaped column, is firmly attached to a vibrating motor (*ca.* 10 to 20 c/s). The shaft of the powerful motor is firmly attached to a centrifugal weight; the vibrating parts are arranged in such a way that the safety requirements are met with. The hub of the vibrator should be 2–3 mm.

After the column has been filled, the ends of the column are plugged with glass wadding, quartz wadding, or inactive asbestos wool, in such a way as to prevent an escape of the packing (dust!) but at the same time to add no noticeable resistance to the gas flow. The plugs should be prevented from slipping by copper gauze (V2A steel mesh may also be used). The column must be packed completely, so that not even the slightest amount of dead space remains.

It is expedient to obtain an exact value in g/ml for the packing density of the material (which is assumed to be introduced into the column without loss) by a preliminary measurement of the volume of the column (with water) and by exact weighing and reweighing of the packing. In this way the reproducibility of the packing may be followed by measurement.

It is especially easy to pack coarsely granular material. A good packing with finely granular material (fine kieselguhr) is very difficult to obtain. At a gas flow of 3 l./hr H_2, an internal diameter of 6 mm and a granule size of 0·2–0·4 mm the pressure drop for a 1 m column should be no greater than 0·1 atm.

Testing the Column before its Incorporation into the Apparatus

Preliminary testing is especially to be recommended when coiled columns are to be incorporated into the apparatus immediately after packing. In this way time can be saved.

Straight columns also should be tested before incorporation. The testing should in each case be related to artificial ageing. The items to be tested are:

1. flow resistance or pressure drop, Δp
2. hold-up time, t_d
3. column performance, n'.

From the results of 1–3 we can calculate

4. performance index.

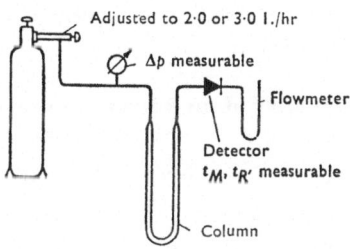

Adjusted to 2·0 or 3·0 l./hr

Δp measurable

Flowmeter

Detector
t_M, t_R measurable

Column

Figure 17. Column testing unit

From the results of 1–4 we can tell whether the packing density has the right character, whether the packing has been carried out homogeneously and whether the column will do what is required of it. The column testing device consists of the units shown in *Figure 17*.

Ageing of the Column Packing

The last residues of the solvent used in impregnation will on the one hand increase the quantity of liquid phase and on the other influence its polarity.

Solvent impurities of low volatility and liquid phase impurities of high volatility must be removed. The extraneous vapours and the traces of air and water in the pores of the support must be eliminated. The layer of liquid which has been introduced must be made homogeneous. All these processes are termed 'primary ageing'. Primary ageing should be complete before the column is incorporated into the apparatus, because otherwise all the above-mentioned volatile substances will get into the detector and contaminate it.

Under no circumstances will a column produce a constant and good base line in the detector during primary ageing. Primary ageing is accelerated by connecting the column to a suitable gas stream but not coupling it up to the detector, leading the carrier gas off for the time being, and working at a temperature slightly above the normal operating temperature.

Every column also undergoes secondary ageing. By this is understood the normal evaporation or the unavoidable thermal alteration of the liquid phase. As every organic liquid has a vapour pressure, the carrier gas flowing over the liquid phase will continually remove a quantity, albeit a small one, of liquid, so that finally only the pure support is left. In continuous use under the most favourable conditions this may take a matter of years, but in many cases it occurs after as few as 1,000 analyses.

Attempts have been made to prevent the process of secondary ageing by saturating the carrier gas stream with the vapour of the liquid phase, but this only produces very disturbing new effects, so that it is better to find a more suitable liquid phase with as low a vapour pressure as possible. Secondary ageing depends primarily on the purity of the carrier gas and the operating temperature.

For liquid phases that consist of mixtures of different components the process of secondary ageing is related to a continual alteration of resolving power, since certain components of the mixture will evaporate more readily than others.

Summary of the Factors Governing the Preparation and Operation of Efficient Columns

The following factors influence the efficiency of a column.

1. *Column length*

The longer a column is, the greater the number of theoretical plates it possesses. A long column, however, requires a long time for an analysis. Under otherwise equal conditions the number of theoretical plates increases (almost) linearly with column length; the retention times of the substances increase in the same way.

Another difficulty is that with increasing column length the resistance to gas flow also increases.

The reasonable limit for packed columns is thus 10 m.

2. Column diameter

Column performance increases with decreasing column diameter. Packed columns normally have an internal diameter of 4 mm, and columns with high separating powers have a diameter of 2–3 mm. However, the load capacity for the substance under investigation decreases with decreasing diameter, so that the column diameter can only be reduced if the detector is sensitive enough to record satisfactorily the reduced quantities of substance.

For packed columns the average load capacity lies at about 3 mg per individual substance for 4 mm columns, and at about 1 mg substance per individual substance for 2–3 mm columns.

3. Gas flow rate and type of carrier gas

For packed columns there is an optimum value for gas flow rate. This value depends on the type of gas being used. Considerably higher flow rates can be used for hydrogen than for gases of higher viscosity, such as nitrogen or helium.

If the flow rate is too low, the column performance is far below the maximum. If the flow rate is too high, the column performance is high enough, but the gas pressure rises sharply.

For packed columns with a diameter of 4 mm (2 mm) the optimum gas flow rate for hydrogen lies in the region of 2·5 l./hr (1 l./hr) when the granule size of the packing is in the range 0·3–0·2 mm (see also p. 170).

The column performance increases with increasing molecular weight of the carrier gas; it is approximately proportional to the root of the carrier gas density.

While hydrogen will give a worse column performance than gases of higher viscosity for high concentrations of liquid phase (30 wt. per cent), this difference is not found for packings with lower concentrations (below 10 wt. per cent). Rapid analyses are therefore carried out with hydrogen at a high flow rate and a low concentration of liquid phase.

4. Nature of the solid support

The column performance is strongly influenced by the type of support. Fine kieselguhr is to be preferred in cases where the pressure drop is not a hindrance. If short analysis times and low pressure drop are required, coarsely granular kieselguhr is to be used. In the separation of strongly polar substances a completely inactive support must be used.

5. Granule size of the support

The column performance increases with falling granule size, but unfortunately the gas pressure and the time of analysis also increase. A compromise is needed here; the optimum value is at a granule size of over 0·1 mm. With increasingly narrow *range* of granule size the column performance increases. The effect of this factor in particular should not be underestimated.

For maximum column performance: 0·125–0·150 mm.
For optimum conditions in practice: 0·250–0·300 mm.

6. *Column packing density*

With increasing packing density the pressure drop along the column increases sharply; above a normal packing density of *ca.* 0·64 g/ml for kieselguhr the column performance remains constant. The finer the granulation of the packing, the greater is the influence of the packing density on column performance. Its values are critical when *less* than the normal packing density is used.

7. *Quantity of liquid phase*

The column performance increases with decreasing amount of liquid phase per amount of support, but the load capacity of the column is considerably reduced.

Minimum value 1 to 5 per cent liquid phase
Maximum value 30 to 40 per cent liquid phase on porous material
Optimum value 10 to 16 per cent liquid phase.

For rapid analyses which still have good column performance a loading of 10 wt. per cent liquid phase is recommended. It should be noted that for such packings the efficiency increases sharply with decrease in loading with the substance under investigation.

8. *Type of liquid phase*

The liquid phase should have as low a viscosity as possible at the operating temperature. Its selectivity must be adjusted to the task in hand, or else even a high number of theoretical plates is no use. Its molecular weight should not be too high. Its vapour pressure must be low; the liquid phase must be able to neutralize the active centres of the support, or else tailing will cause a great decrease in column performance. The effect of the active centres may be eliminated by adding to the liquid phase 1–5 per cent of a strongly polar substance of low volatility (e.g. polyethylene glycol, bases, acids, treatment with chlorsilane, corrosion inhibitor etc.).

9. *Type of liquid film*

The liquid must be spread over the solid support in a homogeneous film of uniform thickness. This will be achieved the more easily, the greater the extent to which the auxiliary solvent used dissolves the liquid phase (no preliminary precipitation) and the more evenly impregnation takes place. The use of the so-called adsorption process for impregnation is greatly to be recommended.

References

10. *Method of introduction of the substance*

The efficiency of the column is strongly influenced by the manner in which the substance is introduced. This factor may be favourably influenced by using the best possible type of sample injector. The aim should be as high a concentration as possible of the substance injected (for gases, no dilution with carrier gas, introduce rapidly; for liquids, vaporize as rapidly as possible). As high a temperature as possible in the sample injector is decisive.

If possible there should be no dead volume between the injection point and the column packing, and likewise there should be none between the column outlet and the detector.

References

1. GOLAY, M. J. E., *Gas Chromatography*, ed. Coates, Noebels, Fagerson, Academic Press Inc., New York, 1958, p. 1.
2. STRUPPE, H. G., Diploma thesis for the Karl-Marx University Leipzig (1958). Faculty of Mathematics and Natural Science.
3. 'Recommendations of the Standardization Committee,' *Vapour Phase Chromatography*, ed. D. H. Desty, Butterworths, 1957, p. xi.
4. JAMES, A. T. and MARTIN, A. J. P., *Biochem, J.*, 1952, **50**, 679.
5. — *Analyst*, 1952, **77**, 915.
6. GLUECKAUF, E., *Trans. Faraday Soc.*, 1955, **51**, 34.
7. CREMER, E. and ROSELIUS, L., *Angew. Chem.*, 1958, **70**, 42.
8. PHILLIPS, C., *Gas Chromatography*, Butterworths, London, 1956.
9. KEULEMANS, A. I. M., *Gas Chromatography*, Reinhold Publ. Corp., New York, 1957.
10. RÖCK, H., *Selected modern separation processes for the purification of organic substances*, Dr. Dietrich Steinkopff, Darmstadt, 1957, pp. 37–135.
11. WIEBE, A. K., *J. Phys. Chem.*, 1956, **60**, 685.
12. DEEMTER, J. J. VAN, ZUIDERWEG, F. J. and KLINKENBERG, K., *Chem. Engng Sci.*, 1956, **5**, 271.
13. SCOTT, R. P. W., *Gas Chromatography*, ed. D. H. Desty, Butterworths, London, 1958, p. 189.
14. MARTIN, A. J. P., *Gas Chromatography*, ed. Coates, Noebels, Fagerson, Academic Press Inc., New York, 1958, p. 237.
15. CVETANOVIC, R. J. and KUTSCHKE, K. O., *Vapour Phase Chromatography*, ed. D. H. Desty, Butterworths, London, 1957, p. 87.
16. JOHNS, T., *Gas Chromatography*, ed. Coates, Noebels, Fagerson, Academic Press Inc., New York, 1958, p. 31.
17. DIMBAT, M., PORTER, P. E. and STROSS, F. H., *Analyt. Chem.*, 1956, **28**, 290.
18. KEULEMANS, A. I. M., and KWANTES, A., *Proc. 4th World Petrol. Congr.*, Rome, 1955, Paper 4 (V/B).
19. SORENSEN, I. B., and SOLTOFT, P. *Acta Chem. Scand.*, 1956, **10**, 1673.
20. CROPPER, F. R. and HEYWOOD, A., *Nature, Lond.*, 1954, **174**, 1063.
21. SIMMONS, M. C. and SNYDER, L. R., *Analyt. Chem.*, 1958, **30**, 32.
22. McFADDEN, W. H., *Analyt. Chem.*, 1958, **30**, 479.
23. KEULEMANS, A. I. M., KWANTES, A. and ZAAL, P., *Analyt. chim. acta*, 1955, **13**, 357.

The Column

24. FREDERICKS, E. M. and BROOKS, F. R., *Analyt. Chem.*, 1956, **28**, 297.
25. ZLATKIS, A., *Analyt. Chem.*, 1958, **30**, 332.
26. CHESHIRE, J. D. and SCOTT, R. P. W., *J. Inst. Petrol.*, 1958, **44**, 74.
27. EGGERTSEN, F. T. and KNIGHT, H. S., *Analyt. Chem.*, 1958, **30**, 15.
28. HRAPIA, H., Dissertation for the Karl-Marx University Leipzig (1959). Faculty of Mathematics and Natural Science.
29. JAMES, A. T. and MARTIN, A. J. P., *Brit. med. Bull.*, 1954, **10**, No. 3, 170.
30. ZLATKIS, A., and LOVELOCK, J. E., *Analyt. Chem.*, 1959, **31**, 620.
31. BAYER, E., Karlsruhe, Private communication.
32. BAKER, W. J., LEE, E. H. and WALL, R. F., *ISA Symposium on Gas Chromatography*. Michigan State University, 1959.
33. HORNING, E. C., MOSCATELLI, E. C. and SWEELEY, C. C., *Chem. and Ind.*, 1959, 751.
34. SZÉKELEY, G., KORMANY, T., RÁCZ, G. and TRAPLY, G., *Periodica Polytech.*, **2**, 1958, 269.
35. GERRARD, W., HAWKES, S. J. and MOONEY, E. F., *Gas Chromatography 1960*, ed. R. P. W. Scott, Butterworths, London, 1960, p. 199.
36. DECORA, A. W. and DINNEEN, G. U., *Analyt. Chem.*, 1960, **32**, 164.
37. WEHRLI, A. and KOVATS, E., *Helv. chim. acta*, 1959, **42**, 2709.
38. PURNELL, J. H. and QUINN, C. P., *Gas Chromatography 1960*, ed. R. P. W. Scott, Butterworths, London, 1960, p. 184.
39. ETTRE, L. S., Paper presented at Pittsburgh Conference on Analytical Chemistry and Applied Spectroscopy, March 1960.
40. LEIBNITZ, E., Leipzig, private communication.

Further literature on the Production of Efficient Columns and the Influence of Particle size

MELLOR, N., *Vapour Phase Chromatography*, ed. D. H. Desty, Butterworths, London, 1957, p. 63.
ADLARD, E. R., *ibid.*, p. 98.
KEULEMANS, A. I. M. and KWANTES, A., *ibid.*, p. 15.

Influence of the Quantity of Stationary Liquid

GRANT, D. W. and VAUGHAN, G. A., *Vapour Phase Chromatography*, ed. D. H. Desty, Butterworths, London, 1957, p. 413.
BROOKS, J., MURRAY, W. and WILLIAMS, A. F., *ibid.*, p. 281.

Influence of the Quantity of Liquid Phase

HARRISON, G. F., *ibid.*, p. 332.
KEPPLER, J. G., DIJKSTRA, G. and SCHOLS, J. A., *ibid.*, p. 222.

2.2. THE CARRIER GAS

In the following section the individual properties of the carrier gases available and the methods for producing, measuring and controlling the gas flow rate are described. The great influence of the purity of the carrier gas and of the specific type of gas are dealt with. The theoretical basis for the influence of the mobile phase in the separation itself was derived in the section on theory. The participation of the carrier gas during the measuring process in the detector will be dealt with in the chapter on detectors.

THE CHOICE of the most suitable carrier gas, together with the correct choice of operating conditions for a given separation, is of great importance. Often too little attention is paid to the purity of the carrier gas. If in addition the measuring and control installations for the gas stream are unsuitable then the conditions for successful operation are not fulfilled, no matter if the best detector, a first-rate recorder and an ideal column are available.

It is a well-known fact that the enormous possibilities of gas chromatography depend on the properties of the mobile phase used; the extremely low viscosity of the carrier gas enables long columns of high efficiency to be used. The elegant methods available for the detection of very small amounts of material in a gas stream enable the quantitative results to be recorded automatically. But this fact indicates an important thing about the carrier gas: *its physical and chemical state should, to the highest possible degree, be kept constant while it is flowing through the column and the detector.*

1. *Hydrogen*

As it has the lowest viscosity and highest thermal conductivity as compared with all other gases, hydrogen is recommended especially for cases when a thermal conductivity cell is to be used as detector and a low column inlet pressure is desired.

Advantages. Hydrogen can very easily be purified. Its maximum thermal conductivity value enables sensitive signals to be obtained when thermal conductivity is to be used as the detection method. Differences in the structures of the substances to be recorded are suppressed to such an extent that, almost always without specific calibration, the chromatogram recorded gives the composition of the sample in weight per cent. Even at high current loading, i.e. high detector sensitivity, the detector is protected, for the average temperature of the measuring wire is comparatively very low in hydrogen. If gas cylinders are not available, pure hydrogen may easily be produced by electrolysis. This has the advantage that the gas flow can be adjusted by means of the electrolytic current, so that the constancy of gas flow is controlled directly by the constancy of the electrolytic current. The pressure and rate controllers can therefore be omitted, since the flow rate can be read off from the ammeter which records the electrolytic current. As in northern countries many high pressure hydrogenation plants obtain high pressure hydrogen directly from the electrolytic cell, it is obvious that the problem of removing the oxygen and compressing the hydrogen is capable of a technically elegant solution.

Disadvantages. It might be supposed that the low viscosity of this gas could only be advantageous, since it meets to a maximum degree the requirements for a mobile phase of low viscosity. But on the other hand this property requires particular care to be taken in the preparation of connecting tubes, valves, taps, and gas tubing because of the danger of gas leakage. In particular, those instruments which use a rubber cap as column closure and have a hypodermic syringe as sample injector (see section on sample injectors) are most liable to go wrong when hydrogen is used as carrier gas.

At operating temperatures below 200°C column packings are generally resistant to hydrogen. Only such a packing as silver nitrate in benzyl cyanide —used for the separation of saturated hydrocarbons from olefins—is reduced at temperatures below 200°C. Sensitive substances in particular, such as conjugated unsaturated fatty acids, can, in the presence of catalytically active metals (column material) or support, undergo chemical change. Generally speaking, however, hydrogen is as a reaction participant far more inert than oxygen. More will be said of this later. The danger of explosions with hydrogen should not be underestimated. Especially where an air thermostat is used to control the temperature, care must be taken in the construction of the instrument to avoid any chance of an oxyhydrogen explosion (washing through with a protective gas)*.

Further, the high diffusivity of hydrogen introduces a whole series of effects which can make their disturbing presence felt in dense column packings, in columns with narrow outlets (microflame detectors) and in exceptionally long columns. The hydrogen flows through the column in an irregular fashion because, due to the great viscosity difference between substance and carrier gas, every substance peak acts as a plug. 'Negative' peaks are formed shortly before the elution of the higher concentrations, and the base line may show gradations behind each component.

Only in countries where helium is not available at a reasonable price can the advantages of hydrogen be set against its disadvantages; otherwise helium is to be preferred.

2. *Nitrogen*

Nitrogen is especially popular as a carrier gas on account of its safety.

Advantages. Nitrogen is readily available at low prices. It may easily be purified. It has no negative effects in the sense that hydrogen has and it also does not require such extreme care in the preparation of the connecting tubes and taps, since in comparison with hydrogen its viscosity is much greater.

---

* The risk of an oxyhydrogen explosion can be almost entirely eliminated by using a simple flow controller, i.e. a metal capillary welded into the pressure reduction valve. The metal capillary, which should be about 50 mm long and have an external diameter of 2–3 mm, and an internal diameter of 0·2–0·3 mm, is compressed with a manual hydraulic press for 20 mm of its length in such a way that a gas flow of 5 l.H$_2$/hr. corresponds to a pressure drop of 1·5–2 atm.

Disadvantages. The thermal conductivity values for nitrogen are so close to those of most organic gases and vapours that where a thermal conductivity detection method is used a specific calibration for quantitative evaluation of the chromatograms is necessary in almost every case. Only when homologous series of organic compounds are being analysed do the peak areas correspond, to a first approximation, to the weight percentages of the homologues. (More of this in Section 3.2 and Volume III.) Nitrogen can certainly be used even at the highest temperatures as a carrier gas, but for temperatures of over 100°C the purity of commercial cylinder nitrogen is just not good enough. It must be freed from oxygen. Normally the oxygen content of commercial nitrogen varies from 2 to 5 per cent vol.* (see under gas purification).

3. Carbon Dioxide

If gases are to be separated for preparative purposes, and if the Janak integral process is being used, then carbon dioxide, which may be absorbed in caustic potash, is used. Its viscosity is even greater than that of nitrogen.

Commercial carbon dioxide, however, is generally contaminated with air (5–10 per cent vol.). Carbon dioxide of greater purity may be obtained by breaking up a block of solid CO_2 (which must be completely compact) rapidly and packing it into a suitable pressure vessel. It is then allowed to blow off, with the valve open, for about 30 minutes, and then the valve is closed and the temperature of the solid CO_2 is regulated. Take care! Not anyone should be allowed to fill a pressure vessel.

Advantages. Carbon dioxide so prepared may be dissolved completely in caustic potash (35 wt. per cent KOH in pure H_2O). Carbon dioxide is perfectly safe and so viscous that the production of gas-tight connecting tubes and valves presents no difficulties. As the carrier gas may be removed quantitatively by absorption, high purity gases may be obtained from mixtures by preparative gas chromatography, which is of interest in the preparation of the noble gases or of radioactive gases.

Disadvantages. Carbon dioxide of a sufficient degree of purity is expensive to prepare. The inlet pressure of carbon dioxide is large in comparison with that of the other carrier gases, which makes high demands on the sample injector.

4. Helium

Apart from its high viscosity, all the desirable qualities for a carrier gas are found in helium. The price of helium is no longer high so that it may be considered for normal applications.

Helium may also be used in instruments equipped with radioactive ionization detectors, but it is better to use argon or neon instead, since otherwise an extraordinarily high degree of purity is required.

* This applies to German commercial nitrogen (technischer Bombenstickstoff). The oxygen content of British commercial nitrogen is *ca.* 0·05 per cent vol.— Translator.

5. *Argon*

Advantages. Its complete chemical inertness recommends the use of argon as carrier gas for very high temperature separations, in which a β-ray ionization detector or a microflame ionization detector is used. But also at normal operating temperatures the use of argon is advantageous, since from its method of preparation it contains practically no oxygen and thus does not require purification, although occasionally admixed methane can cause trouble. The price is relatively low, so that in the near future the use of argon is likely to increase considerably.

6. *Air and Oxygen*

When the substance to be separated is burnt and is to be recorded as CO_2 or H_2 (obtained from the H_2O of the combustion products by reduction over Fe—see section on detectors) or recorded with the microflame detector, and when the operating temperature is less than 100°C, these gases may also be used.

7. *Other Gases*

Always when traces of a gas or vapour are to be directly detected in another gas by the thermal conductivity method, it is worth using the trace gases themselves (in as high a state of purity as possible) as carrier gas. This simplifies the direct recording of the trace gases. Even for special cases like these, the basic requirement for the mobile phase must be met; the carrier gas must not be soluble in the stationary phase. This requirement is satisfactorily met by the other gases described above.

Gas Purification

All commercial grade gases contain oxygen, water, oil vapours, etc. Hydrogen from hydrogenation processes is often contaminated with lower hydrocarbons.

We can distinguish two types of impurity: the first type are those that dissolve in the stationary phase, i.e. in the liquid phase, or load the adsorbent in gas solid chromatography. Even if these impurities are only traces, it must be considered that these traces are continually being added to the separating column. This leads finally to a saturation of the column packing with impurities. Now the degree of saturation is temperature dependent. Every increase in temperature leads to a sudden and now concentrated emission of impurities within the column packing. On the one hand this causes a continuous variation in the measurable gas concentration; signals are thus obtained from the detector. On the other hand, of course, the efficiency of the stationary phase varies. This can lead to reversible or irreversible alterations of the stationary phase. Towards many of the impurities in the carrier gas the column acts like a sponge, which takes up impurities as the temperature falls and gives them off as it rises. The base line thus becomes strongly temperature dependent, and is particularly unsatisfactory when techniques other than the isothermal ones are used (e.g. for progressive or programmed temperature control of the column). The quality of the quantitative signal suffers under such conditions.

The negative effect of impurities of the second type is also greatly underestimated. The substance generally concerned here is oxygen. At operating temperatures of over 100°C, hydrocarbons, some silicone oils and a whole series of other liquid phases with long aliphatic chains are liable to undergo the so-called atmospheric oxidation. This oxidation by molecular oxygen even takes place when the concentration of oxygen in the gas is below 1 per cent vol. Such an oxidation process can occur immediately; generally, however, the so-called oxidation inhibitors (e.g. very small traces of phenols, amines, etc.) can prevent the onset of oxidation for a considerable time. But once it has started the first formed oxidation products have a catalytic effect, and the process will even continue when hydrogen containing only 1 per cent oxygen is used as carrier gas at temperatures of over 100°C. A column packing thus oxidized has considerably different properties, becomes strongly temperature sensitive and often reacts with the substances to be separated. The aggressive degradation products (CO, formic acid, aldehydes and unsaturated compounds) decompose in the detector or on heated metal surfaces. In the end the entire apparatus must be dismantled and thoroughly cleaned, before the fresh column packing needed is introduced.

Methods of Purification

Silica gel or active carbon at −180°C remove all impurities except oxygen. The purifier should cause no additional pressure drop, and thus the cross section must be correspondingly large (e.g. 5 cm diameter, 30 cm high, cooling traps made of difficultly fusible glass and placed in Dewar flasks).

The surest way of removing oxygen is to pass the gas through a V2A steel tube packed with copper and heated to 450°C: the copper is obtained from reduction of copper oxide (for elemental analysis)*. Any oxygen in the hydrogen is here burnt to form water; the copper remains constantly active. The water formed must either be frozen out or else removed by a phosphorus pentoxide drying tower.

More suitable than P_2O_5 is a drying column which consists of a 20 cm long layer of silica gel (containing a moisture indicator, e.g. $Co(NO_2)_3$ followed by a layer of molecular sieve of type 3A or 4A dried at 400°C. This packing is also effective at high temperatures and can be easily regenerated. It can be used as often as desired and remains granular, and CO_2 is also absorbed by it.

Water can also react with powdered iron at 750°C to give hydrogen and iron oxide. It is best for the gas connections to be of metal, so that the oxygen absorber can be heated to 450°C with safety. So that the heat evolved cannot do any damage, air coolers are mounted at both ends of the V2A tube (see *Figure 18*). The purifier can easily be regenerated in the normal way. Copper oxide can easily be reduced by hydrogen at 450°C. It is, however, recommended that the purifier should be filled with hydrogen while cold, and should only be heated for the reduction while hydrogen is flowing through it. The maximum flow rate should not exceed 10 l./hr, or else local over-heating from the heat of combustion can easily cause melting of the copper, with all the undesirable accompaniments. The completion of the reaction

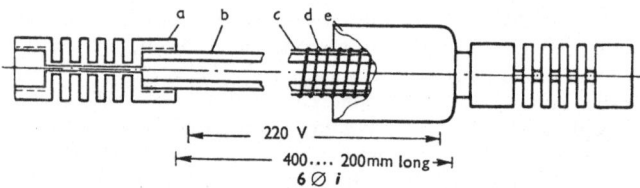

Figure 18. V2A steel gas purifier packed with copper wire. The heating coil must produce an internal temperature of 750°C.

a = brass plated with nickel on the   c = layer of asbestos paper
    outside and gold on the inside   d = edge winding
b = V2A steel, tube 8 × 1   e = good insulation

should also be followed, to prevent a failure of the purifier leading to the destruction of valuable column packings.

Reich and Kapenekas (*Ind. Engng Chem.*, 1957, **49,** 869) used an aqueous solution of sodium dithionite, sodium phosphate and zinc sulphate for the purification of nitrogen. In this way they were able to keep the residual oxygen content down to 3×10^{-2} per cent. For high temperature analysis the maximum O_2 content is <0.5 p.p.m.

The regeneration of active carbon and silica gel absorbents is also suffici-ently well known; it is expedient to carry out this regeneration in the actual cooling trap vessel. One is thus saved the bother of assembling the apparatus and making everything gas tight again.

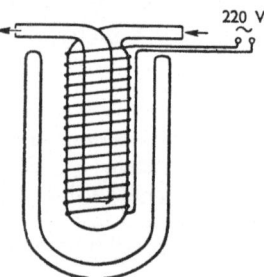

Figure 19. Gas purifier packed with a mixture of active carbon and molecular sieve. Arranged for cooling and supplied with heating system for regeneration purposes.

Oxygen is removed from hydrogen in the cold at 20°C with *ca.* 30 g palladium asbestos at a gas flow rate of 2 l./hr.

The absorber should be heated by means of an electrical resistance heating coil around the vessel (see *Figure 19*).

Control and Measurement of the Carrier Gas Flow Rate

Several of the different types of detectors are more or less sensitive to alterations in the flow rate of the carrier gas. Such alterations cause very troublesome displacements of the base line. In such cases it is no longer possible to use automatic instruments or instruments with integrators for evaluation. But even if the detectors used are not sensitive to fluctuations in carrier gas flow rate, quantitative recording is endangered, for at increased flow rate the same quantity of the substance to be recorded passes through the detector in a shorter time, which leads to a change in the mass transport per unit time, and thus to a decrease in the deflection/time integral and with it to incorrect recording of the quantity. Finally, fluctuations in the carrier gas flow rate affect the separating conditions of the column, which leads to a change in the column characteristics for qualitative analysis (altered retention time, altered number of theoretical plates, etc.).

The carrier gas velocity, or, to put it more correctly, the carrier gas flow rate, must therefore be kept constant. For quantitative analysis where an accuracy of ± 1 per cent is required, the flow rate may not fluctuate by more than ± 0·2 per cent. This requirement is indeed difficult to meet.

Disturbances of Constant Flow Rate

The column represents an obstacle to the flow of the carrier gas. This flow resistance depends primarily on the column temperature (viscosity and volume of gas alter with temperature).

Secondly, the flow resistance in a packed column is influenced by variations in the density and form of packing of the support (ageing). The composition of the carrier gas alters continuously during a gas chromatographic analysis, since the nature, volume and number of the individual components of the sample have a considerable influence on the total properties of the carrier gas. The individual substances move as more or less viscous 'gas plugs' through the free volume of the column. More and more 'plugs' leave the column; at the end of the analysis pure gas once more flows along the entire length of the column. Thus the flow resistance varies during an analysis. But a constant amount of gas can only flow through the detector if the ratio of pressure drop along the column to the column flow resistance is kept constant (analogously to Ohm's law; to a first approximation $I = U/R$, and flow rate = pressure drop/flow resistance).

From this we can see that it is absolutely essential for the column to be supplied with a flow of gas which is kept constant even before it enters the column, as the column flow resistance alters continuously and gradually during an analysis, even if the external operating conditions remain constant. It is not possible to achieve a constant flow rate simply by producing a constant pressure before the column inlet. To understand the controlling conditions it is useful to recall the relationship (here given approximately) between the pressure drop at a gas resistance and the flowing volume:

$$V = \frac{(p_i^2 - p_o^2) \cdot \pi \cdot r^4 \cdot t}{p_o \cdot 8 \cdot \eta \cdot L}$$

$(p_i^2 - p_o^2)$ = pressure drop; r = inner radius of the capillary or an equivalent value of the gas cross sectional area of the column; η = viscosity of the gas; L = length of the capillary or column (valid only for very short columns which, however, are relatively long considering that they are capillaries).

If this is applied to the conditions of a column acting as a flow resistance there are, of course, a whole series of further factors, which are connected with the interrupted laminarity of flow and other effects. But to a first approximation we can use the above relationship. From this we can see that the gas flow rate through a flow resistance depends on the pressure gradient. If we allow the carrier gas to escape into the atmosphere and maintain a constant pressure at the column inlet by means of a pressure controller, the pressure drop along the column, and with it the flow rate, is constant. But during the course of the analysis the gas composition, i.e. the effective average viscosity, varies. In this way the gas flow rate can fluctuate even when there is a constant pressure at the column inlet, as has already been established.

Production of a Constant Gas Flow Rate before the Column Inlet

A very large pressure drop along a capillary will enforce a constant gas flow even if there are small pressure variations in the gas path after the capillary; these variations must, however, be small in comparison with the inlet pressure before the capillary. We may write:

$$V = (P_1 - p_i) \cdot k + (p_i - p_o) \cdot k'$$

V = constant if $(P_1 - p_i) \cdot k$ is considerably greater than $(p_i - p_o) \cdot k'$ and if P_1 is considerably greater than p_i'

P_1 = primary pressure before the capillary; k = capillary constant;

p_i = column inlet pressure; p_o = column outlet pressure; k' = column constant.

Thus if the direct cylinder pressure (e.g. 100 atm) is used as the primary pressure P_1 and reduced by means of a metal capillary or a very good needle valve to p_i (e.g. 1 atm) then pressure fluctuations within the column (e.g. $\pm$ 0·05 atm) will only have a very slight effect (= $\pm$ 0·05 per cent).

In this way the flow rate can be kept constant even if the temperature of the column is changed, because only inlet pressure changes of about 1 per cent per degree occur. In other words, a column which has a pressure drop of 0·1 atm gauge pressure for 3 l. carrier gas/hr at 20°C will have a pressure drop at 220°C of 0·3 atm gauge pressure. From this we get, to a first approximation, a change in flow rate of 0·2 per cent, corresponding to 6 c.c. gas/hr at 3 l./hr if the primary pressure is 100 atm gauge pressure. For a primary pressure of 40 atm gauge pressure a change of 1·2 per cent occurs under the same conditions.

For practical purposes, instead of a short capillary, a precision needle valve is used. This valve may either be connected directly to the gas cylinder or supplied with gas at a pressure which is kept constant and as high as possible. It is even more advantageous if an especially long metal capillary can be attached to the cylinder. If the temperature of this capillary (which should be about 10 m long) is controlled, then the flow rate can also be varied by varying the temperature.

It should, however, be realized that the above considerations are only valid if the gas emerging from the capillary or needle valve can flow on unhindered. If the gas flow after a needle valve is blocked (e.g. by a tap) the gas pressure at this point rises until it reaches the primary pressure before the capillary. For apparatus where all the connections are made of metal this can lead finally to dangerous excess pressures. For this reason every capillary flow resistance must be fitted with an excess pressure safety device. As a last resort this may be achieved by the use of a short length of vacuum rubber tube (maximum pressure 3–4 atm gauge pressure); it is better to use an officially approved safety device.

For flow rate control a pressure controller with a high outlet pressure (at least 10 atm gauge pressure) is needed as a connecting element for the high precision needle valve.

Needle valves are important pieces of equipment, and are often used in gas chromatography; the properties required of such valves will therefore be described.

In principle the needle valve takes the place of a capillary. It has the advantage of possessing a channel capable of wide variation. If it is remembered that the flow rate varies with the fourth power of the internal diameter of the capillary it may be understood that the bore of this channel must be capable of very fine adjustment. This requires a very slender needle (low taper) with a precise seat at its tip. The needle must be rotatable by a very fine thread and should be highly polished. The needle and needle seat must be made of hard materials; plastic insets which are supposed to prevent the needle jamming not only give way at such moments but also over a very long period of time change the fine free cross section of the bore of the channel. For example, the cross-sectional area of a gas passage in a needle valve designed to allow a flow of not more than 3 l. gas/hr from a hydrogen cylinder at 150 atm gauge pressure is only a few thousandths of a square millimetre. For this reason the gas entering the valve should first have been purified by passage through a very fine filter. The needle valve may not be exposed to any temperature fluctuations. One might go so far as to require a temperature control system for the needle valve.

The needle valve must naturally be completely gas tight externally. As, however, the needle is rotatable, the driving shaft must be sealed off from the outside with a gland. Completely gas tight glands can easily be prepared from teflon or teflon powder mixed with asbestos. Teflon flows under pressure and is self-lubricating. Other materials such as leather, rubber, grease with asbestos twine, polyethylene, polyvinyl chloride, and other substances used for glands where the requirements for gas tightness are not very stringent have not proved effective for the present purpose in continuous use.

An interesting and constructive solution to the need for a device for keeping the flow rate constant is given by Knox (Knox, J. H., *Chem. & Ind.*, 1959, 1085). In principle the glass apparatus shown in *Figure 20* corresponds to the Sunvic[1] flow controller. The instrument is supplied at A with gas pressure 200 mm higher than the highest column inlet pressure which may occur, e.g. in a gas chromathermographic analysis, from the temperature increase at the column inlet at constant gas flow. The instrument has an adjustable needle valve N, a rubber stopper G (freed from sticking with

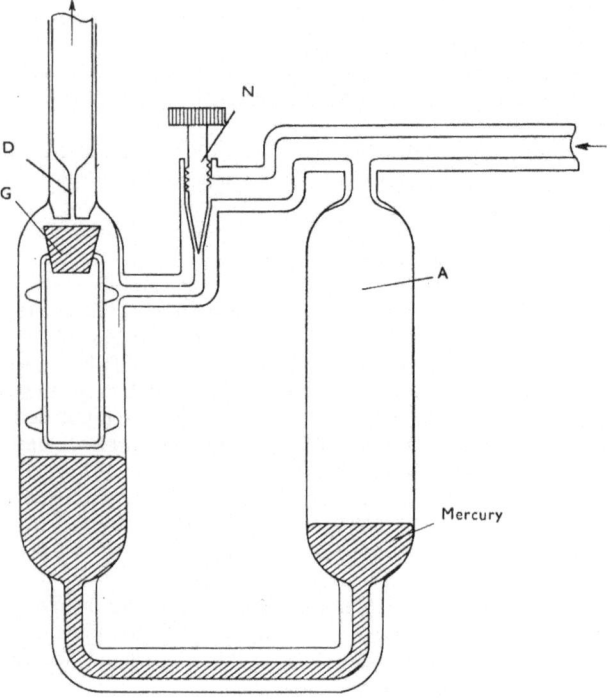

Figure 20. Knox gas flow rate controller.

french chalk), which is supported on a glass float by mercury, and a constriction D, the underneath of which is ground somewhat obliquely. This instrument keeps the gas flow constant within 1 per cent at 100 ml/min, when the column temperature rises from 20 to 120°C.

Pressure Controllers

A pressure controller should be placed in the flow path before the needle valve or capillary. A gas cylinder will, of course, provide a practically constant pressure source over a more or less long period of time, since normally only a little gas is removed. For example, a nitrogen cylinder of normal size will at a gas consumption of 3 l./hr last for approximately 1 to 2 months, assuming that no leaks occur. Pressure constancy during an analysis may thus be taken for granted. Over longer periods of time, however, an adjustment of the needle valve is necessary. This may be avoided if a pressure controller is placed before the needle valve.

Normal pressure reduction valves which give a 'constant' outlet pressure in adjustable values are not suitable for the low gas flow rates needed for gas chromatography. It is known that pressure reduction valves control the pressure on the outlet side by means of membrane controlled ball or cone valve seals.

Figure 21 shows how a gas flow controller, e.g. as supplied by Sunvic Controls Ltd., London, can be converted into an excellent pressure controller. A connecting membrane with constant gas pressure, e.g. from a gas tight buffer vessel, is set up.

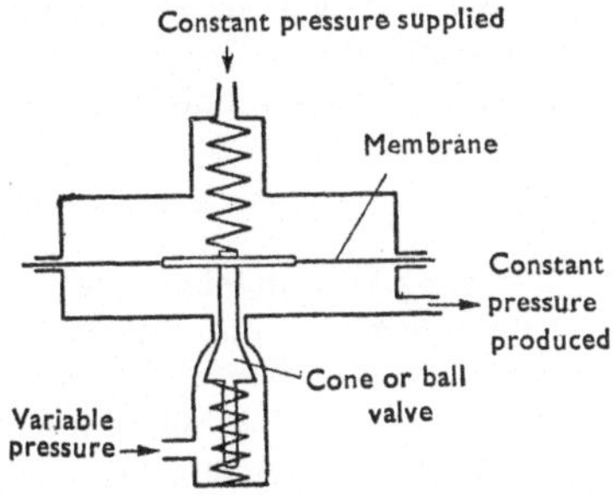

Constant pressure supplied

Membrane

Constant pressure produced

Cone or ball valve

Variable pressure

Figure 21. Plan of the Sunvic flow through controller converted for pressure control.

Pressure controllers that remove excess pressure by blowing off are generally unsuitable for gas chromatography. (Helium, argon and hydrogen are too valuable or too dangerous, while high purity nitrogen is too expensive.)

Other gas flow controllers are described by Skrokov, M. R., *Power*, 1957, No. 2, 101; Kulcsar, G. J., Kulcsar-Novakova, M., *Acad. rep. populare Romîne, Filiala Cluj, Studii cercertâri chim*, 1956, **7**, No. 1–4, 119; James, D. H., Phillips, C. S. G., *J. sci. Instrum.*, 1952, **29**, 362.

Gas Tightness

All the trouble taken to provide constant pressure and constant gas flow will be in vain if there are leaks in the gas tubing and connections. Such leaks generally have the property of varying the gas flow in a completely irregular fashion. This leads to irregular disturbances. Further, for gas chromatographs using hydrogen, the explosion risk is increased.

It is expedient to test the following parts of the apparatus with an aqueous (5 per cent) solution of Nekal: main valve of the cylinder, pressure control valve with manometer connection, needle valve and gland tubing connections and the tubing itself over its whole length (!), sample injector and head of the column. It should be noted that the really large leaks will not be shown up by the Nekal, since the consistent film necessary to give the foam will not be formed. It is expedient not only to paint the Nekal on as solution but first to form it into a stiff foam. Further, it is necessary to check the detector more frequently, since temperature fluctuations can cause slight fissures etc., even in a compact metal block.

For leak testing, hydrogen under a pressure of 1 to 2 atm gauge pressure should always be used, since its low viscosity will enable it to pass through even the finest holes. This method of testing is especially advantageous for instruments that work under vacuum. It is possible to do away with the rather inconvenient Nekal process by using special very sensitive thermal conductivity cells—the so-called leak detectors. Such detectors usually give an acoustical signal when the detector probe comes to a leak. For this method the whole apparatus to be tested is placed under vacuum, the measuring cell is placed at the outlet to the suction point; the detector probe consists of a hydrogen source with a fine nozzle through which a little hydrogen continually flows. If the hydrogen is drawn into the apparatus as a result of a drop in pressure it flows over the measuring cell and produces a signal.

Flow Rate Meters

There are several methods for measuring the gas flow rate, but they all agree upon one point; the reading will be incorrect if the gas laws are not taken into account. The exact measurement of the gas flow rate is important for controlling the operation of the instrument and for obtaining retention values for qualitative analysis.

To measure the flow rate correctly is not an easy task. First must be decided which method of measurement is to be used and at which point in the gas path the measurement is to be made. The point and method of measurement have a partial influence on one another.

Point of Measurement

For continuous observation the point of measurement should be just before the sample injector; for discontinuous measurements which can be carried out with considerably more certainty, the point of free gas emergence after the detector should be chosen.

Methods of Measurement

1. Rotameter

A rotating cone or a ball is placed in a tube which is slightly widened at the top to form a cone. The gas flow produces a dynamic pressure which keeps the object, which is free to move in the gas stream, at a constant height within the conical tube. The height of suspension is a measure of the dynamic pressure. The scale can be divided on a linear basis.

Advantages. The instrument can stand up to overloading without additional help.

Disadvantages. The signal is pressure-dependent. Rotameters must be inserted in the gas path before the sample injector, as they are very sensitive to contamination (by substances in the sample, or by vaporized stationary phase). Rotameters are dependent on the type of gas used.

2. *Capillary Meter*

According to the rule $V/t = (p_i - p_o) \cdot k$, a pressure drop linearly dependent on flow rate will prevail along a short capillary. The pressure drop is
measured (generally by a water or oil column) and is a measure for the flow
rate. The scale is linear. The author has used the type of capillary meter
shown in *Figure 22* with success. It has the advantage that the meter can also
be overloaded.

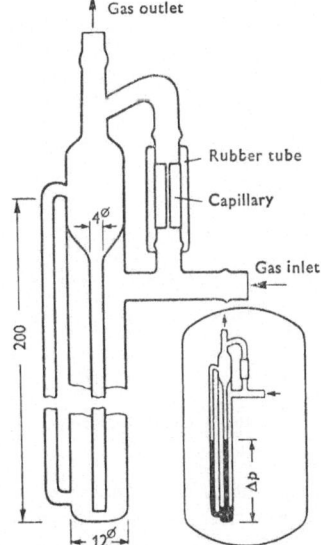

Figure 22. Capillary flowmeter.

The signal given by the meter is pressure-dependent (see under *rota-
meter*), and also depends on the type of gas used. For choice the capillary
meter should only be inserted in the outlet gases for discontinuous measurements. Normally it is placed in front of the sample injector.

3. *Hot Wire Meter*

The flowing gases can remove heat energy from a hot wire. If the wire is
very hot (white heat) the heat loss becomes strongly flow rate dependent.
The flow rate can be read off on a milliammeter. The scale is non-linear.
Hot wire flowmeters are strongly dependent on the type of gas used.

Disadvantages. If hydrogen is used as carrier gas there is the danger of
an oxyhydrogen explosion if oxygen diffusing in from the atmosphere pene-
trates as far as the hot wire flowmeter.

4. *Soap-film Flowmeters*

This instrument is used especially in British and American laboratories,
and is very simple. The carrier gas is bubbled for a short time through a

solution of potassium oleate. The soap films thus formed rise up a tube. The rate of ascent of the soap films is an exact measure of the gas flow rate (it is a direct volume rate measurement). The measuring process is independent of the type of gas used, simple, and cheap. The instrument is connected to

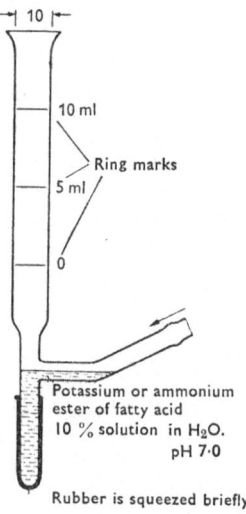

Figure 23. Soap film flowmeter.

the gas outlet, and is best used only discontinuously. This is managed by arranging that the soap solution shall only be pressed into the gas path for short periods by pressure on a rubber tube. The passage of a soap film between two rings which contain a given volume (e.g. 5 or 10 ml) is timed with a stopwatch. Derivation of result: e.g. 10 ml in 16 sec corresponds to 2·2 l./hr. (See *Figure 23*).

2.3. THE SAMPLE INJECTOR

DIFFERENT separations require different methods for introducing the substance to be separated into the gas chromatograph. Thus it is understandable that there are great differences between the methods for the introduction of liquid and gaseous samples, and the method chosen also depends on whether it is a question of routine tests or of jobs that are constantly changing. In all cases, however, there are three conditions which must be fulfilled in the introduction of the substance:

1. The sample should be introduced into the column packing as vapour in the smallest possible space and in as short a time as possible.
2. It should not upset the equilibrium conditions of the column in the process.
3. Both the quantity of substance introduced and the manner in which it is introduced (the prevailing conditions such as temperature, pressure, rate and energy conversion) must be reproducible with as high a degree of precision as possible.

These three conditions are not easy to fulfil, especially when one remembers that for analytical purposes the efficiency of the method is increased when the quantity of substance is decreased. The usual amount of substance is, for a gas, from 0·01 to 5 c.c., which in the case of ethane is 20 γ to 6,000 γ per analysis. For fluids and solids it is normal to use 20 γ to 20 mg substance for an analysis.

Recent development in analytical gas chromatography has, however, been in the direction of very high efficiency column performance, a direction which will be generally welcomed by analysts.

For industrial applications it is required that a gas chromatograph should work automatically with maximum precision and minimum attendance. In this case the requirements made of the sample injection system become a problem only partially capable of solution.

The foundations for the three requirements for sample injection are as follows:

1. The substance is introduced as gas or vapour into the carrier gas stream, which takes it to the first theoretical plate where the separating process begins. The time between the beginning and the end of the introduction process and the reaching of the first plate determines, together with the linear carrier gas flow rate, the minimum or starting width of the peak which will later appear at the end of the column. Certainly the 'starting width' b_0 has been widened considerably as a result of diffusion by the time the end of the separation is reached. The total width is made up as follows: $b_{\text{total}} = b_0 + b_{\text{diff}}$. (the broadening due to diffusion and separation), where $b_{\text{diff}} = D_i \cdot t_i$ (D_i = rate of diffusion in the appropriate direction, t_i = retention time of substance i).

But as $b_{\text{diff.}}$ is a value dependent on the substance, retention time and the operating conditions of the column, it is b_0 which is the variable, influenced by the separating powers of the column (always in a negative sense) the more that the latter differs from $b_{\text{diff.}}$. But the broader the bands, the lesser is the degree of resolution between neighbouring components and the more difficult it becomes to carry out a quantitative evaluation. The applicability of automatic integrating instruments is decreased, as the latter generally need a length of base line between the individual components. The entire process can also be regarded as a decrease in the number of theoretical plates in the column. The time from the beginning to the end of the injection process should not be regarded simply as the time taken by the mechanical process of introducing the substance into the column head, but rather the whole time needed until the last remnant of the substance has reached the first plate of the column. This includes both the sample vaporization time and the time taken to wash the substance through the so-called 'dead column head volume' between the point of injection and the surface of the first theoretical plate.

An example may make the latter clear. Suppose that between the point of injection and the surface of the column packing there is a free volume of 10 c.c. filled with carrier gas. If the gas flow rate is 3,000 ml/hr = 50 ml/min, then it will take 1·4 min (gas changed 7 times) to wash all the product (introduced as gas) out of this volume. In a recorder with a chart speed of 60 cm/hr a peak width of 1·4 cm will already have been recorded, which must be counted as b_0. (See *Figure 24*.)

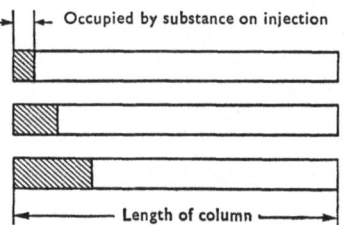

Occupied by substance on injection

Length of column

Figure 24. Decrease in effective column length and spreading of the initial width of a band when the sample is injected too slowly or the sample volume is too great.

From this we can see that the sample injector is not to be regarded merely as the injection device itself but rather as the entire section of the apparatus from the column head to the surface of the column packing, and should be so constructed that the three requirements given above are met to the fullest possible degree.

The influence of the injection technique is critical in the injection of liquids or solids. Here the entire substance must first be converted into vapour, which requires evaporation (or melting) energy which must be instantly available.

The important thing here is to have suitable heat transfer resistances and heat capacities at a temperature which, in spite of this, is as similar as possible to the column temperature. A sample injector that is too cold can falsify the results to a very marked degree. The correct temperature is that at which a drop of the substance evaporates with a hiss.

2. The limited capacity of the column has already been noted. It determines the maximum amount of substance that may be injected if the column performance is to be fully maintained.

The maximum amount may be regarded practically as the amount which can be taken up in the region of a theoretical plate without disturbance of the separation process.

In spite of this a column can normally be loaded with n times the maximum amount, provided that there are n clearly separable components in the mixture (load capacity, *Figure 25*).

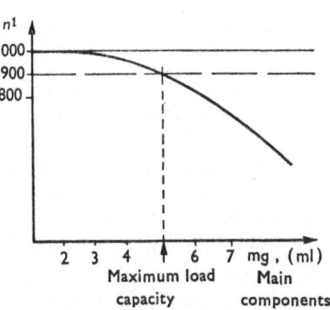

Figure 25. Decrease in the column performance with too great a sample volume, as a definition of load capacity.

Requirement 2 may be regarded as most closely fulfilled if the maximum amount is firstly not exceeded, and secondly not supplied to the first and following plates at a rate at which, on account of the prevailing carrier gas flow rate and rate of diffusion, they are unable to accept it. The column performance and the extent to which the results can be evaluated are negatively influenced if requirement 2 is ignored. The limited rate of supply can only be achieved as a compromise with requirement 1; however, this factor is not so decisive. Sudden pressure surges during injection should, however, be avoided.

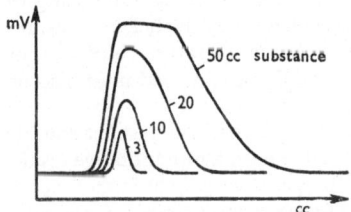

Figure 26. Peak form and relative position when too great a sample volume is taken (from Keulemans, *Gas Chromatography*, Reinhold Publ. Corp., New York, 1959).

Not only the quantity but also, in some cases, the concentration of the substance to be injected in admixture with other substances influences the column performance and the peak form in the chromatogram (see *Figure*

27). As a rule it is better to introduce the substance in maximum concentration, i.e. if possible without dilution by a gas (e.g. methane diluted with nitrogen) or a liquid (e.g. naphthalene diluted with benzene). This dilution cannot always be avoided. Porter, Deal and Stross[1] have investigated these influences and formulated them mathematically.

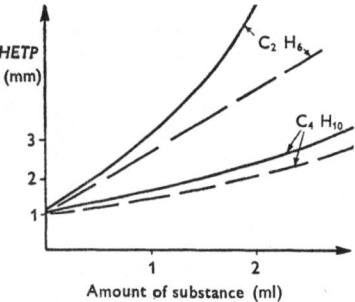

Figure 27. Decrease in height equivalent to a theoretical plate (decrease in column performance) on injecting too highly diluted a sample. ———— 5 per cent substance in N_2. — — — 100 per cent substance.

3. Even though there are methods of evaluation in which it is possible to introduce an unknown amount of substance into the gas chromatograph and still obtain a quantitative result (see Section 3·2), for many tasks the quantitative measurement of the quantity injected is unavoidable. For gases this is relatively simple. A temperature-controlled container of absolutely constant volume is filled with the gas under investigation, pressure and temperature fluctuations being kept so small that they have no noticeable influence on the quantity of substance (influence of less than $\pm$ 0·5 per cent). After it has been filled, the container is connected to the carrier gas stream, and the gas is washed out into the column. For liquids it is practically impossible to carry out such a process. The container volume would have to be smaller by 3–4 powers of ten. Processes suitable for gases fail when applied to liquids because of the adhesion of liquids to the vessel walls, unless adhesion is prevented by total vaporization.

The temperature at the injection point affects the peak height of the substance in the gas chromatogram (cf. Harrison[2]). The peak area is certainly independent of the injection temperature (so integrators will give correct results), but if the results are to be evaluated by the height measurement process it is essential to keep the injection temperature constant. For the analysis of liquids this means that the total amount used must also be kept constant (heat of evaporation!).

Requirement 3 does not need any further theoretical justification. It follows from requirements 1 and 2, and from the need for quantitative results of high precision in both analytical and preparative gas chromatography.

For the physico-chemical application of gas chromatography (determination of partition coefficients, activities, etc.) an optimum injection system is the first prerequisite.

In short we can say: the type and manner of sample injection have an important influence on the gas chromatographic result. The quantity of substance should be as small as possible. It should be introduced quickly and rapidly vaporized if the sample is liquid or solid. The temperature at the injection point should be controllable and constant.

For quantitative analysis all the prevailing conditions must be reproducible with a maximum degree of accuracy.

Sample Injectors for Gases

1. *Hypodermic Syringes*

The gas sample is drawn up into a medical hypodermic syringe, and the needle is pushed through a rubber cap which covers the top of the column. The gas is expelled into the carrier gas stream, and the very fine needle of the syringe is withdrawn from the rubber cap, which seals itself off again. (Ray[3], see *Figure 28* a and b.)

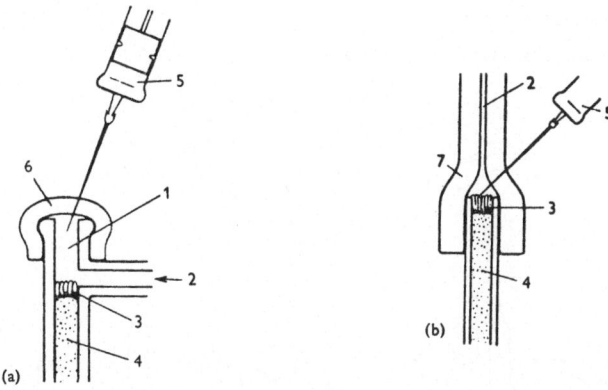

Figure 28. Use of hypodermic syringe to introduce sample (a) through plug or rubber cap; (b) into very simple laboratory instrument with open column.

Variations of the Method. The column packing reaches up to the top and the carrier gas does not enter from the side but directly through the column head. The gas tubing consists of vacuum rubber tubing. The needle is pushed obliquely through the vacuum tubing and the substance is expelled into the column shortly before the start of the column packing.

Advantages. The process enables the injection of any volume required from 0·001 to 50 c.c. It is simple and capable of variation.

Disadvantages. A reproducibility of ± 1 per cent can only be obtained by experienced workers. Ordinary rubber caps need to be changed after they have been used 20 or 30 times, special rubber (silicone rubber) caps can be used up to about 60 times, provided that very fine needles—the so-called 20 cannula—are used.

The syringes are seldom completely gastight (they should be tested frequently by depressing the air-filled syringe under water, the needle being stuck into a rubber stopper). Syringes with glass pistons are advantageous. Different types of gases require different types of lubricant for the piston guide:

hydrocarbon gases: glycerine or phosphoric acid, starch solution
strongly polar gases: hydrocarbon-vaseline, rubber-oil mixtures.

The dead volume of an ordinary syringe is large and makes an exact injection difficult because of the ease with which air can diffuse in. It is necessary to rinse it out several times. By using a metal nozzle the dead volume can be considerably reduced. The trouble taken to control the gas flow rate can be rendered partially ineffective by the use of a rubber cap that is not entirely gastight (having been used a great many times). The use of a syringe is not recommended for integrating instruments.

Assessment. Limited application for quantitative analysis, replaceable by better methods, not capable of automation.

2. By-Pass Systems

The details may be seen in *Figure 29*.

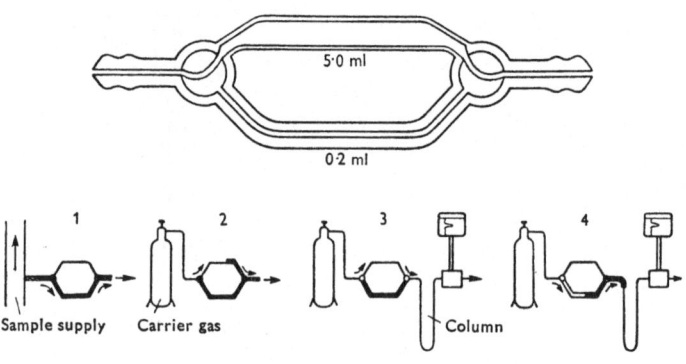

Figure 29. Gas by-pass system with Tschako taps.

1 = filling
2 = preparation (purging)
3 = connection to carrier gas supply and column; air which has entered is washed out
4 = injection

The system should at no point have an internal diameter of more than 6 mm, because in this way the gas sample is introduced into the column packing as a 'plug' without dilution and 'spreading', supposing that the rest of the gas path has a diameter of 1–2 mm. For tubes of too great a bore the rule that the amount of carrier gas required for rinsing out is seven times the sample volume becomes increasingly true.

A more elegant process is that recommended by van de Craats[4], shown in *Figure 30*, which enables the by-pass system to be filled, or even exchanged, and injected without interrupting the gas stream.

Advantages. If the gas laws are taken into consideration the by-pass process is very accurate. It may be recommended for quantitative analysis, and is well suited to routine operations. Aggressive gases and vapours can be injected by this method, provided that suitable lubricants are chosen for the stopcocks.

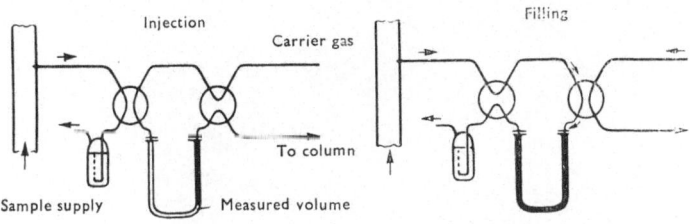

Figure 30. Gas by-pass system with double Tschako taps (from de Craats in Keulemans, *Gas Chromatography*, Reinhold Publ. Corp., New York, 1959).

Disadvantages. Stopcocks which are in frequent use need careful attention. In this connection, the use of either too much or too little lubricant produces the same difficulties. Stopcocks quickly develop leaks. The by-pass system is also normally working under excess pressure when it is connected to the gas stream.

Assessment. When correctly maintained it is very accurate. $\sim \pm 0 \cdot 1$ per cent. Difficult to make automatic.

3. *Gas Sampling Valves*

The details may be seen in *Figure 31.*

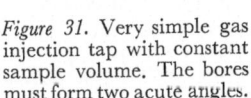

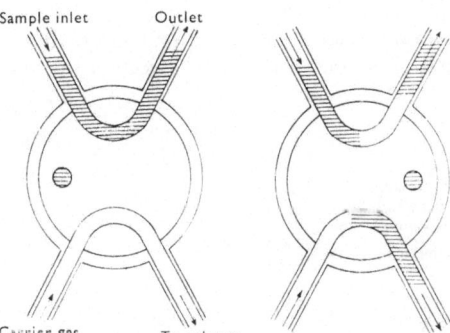

Figure 31. Very simple gas injection tap with constant sample volume. The bores must form two acute angles.

Advantages. Cheap, suitable for automatic operation, recommended for routine analysis, provided that there is a lubricant available which is suitable for the product under investigation. Where necessary the glass stopcock should be enclosed in a teflon casing.

Disadvantages. Only a constant volume may be injected. Maintenance is necessary. Can only be operated at constant temperature.

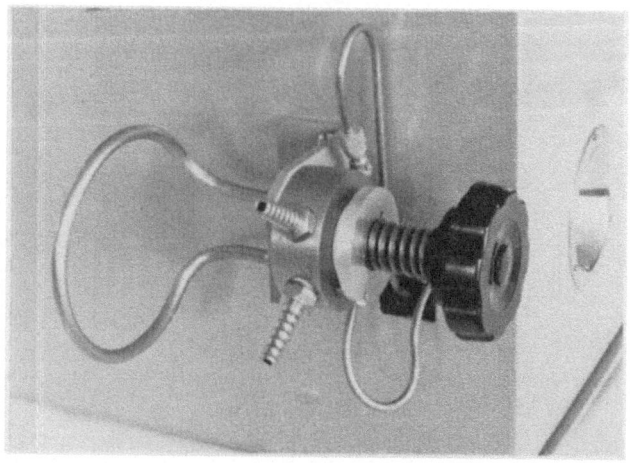

Figure 32. Metal gas injection tap manufactured by Perkin Elmer GmbH Bodenseewerk. (Variable volume, plate principle.) High precision steel surfaces can also be arranged to work without lubricants for gas and vapour injection. Such sample injectors work at any desired temperature up to 400°C.

Assessment. Approved for semi-automatic and automatic industrial control instruments for continuous use. Cheap. (Kögler[5].)

An industrial application is shown in *Figure 32.*

4. Pressure Injection

In the other processes the amount of substance is determined by the variation of volume. In pressure injection the volume is kept constant but the

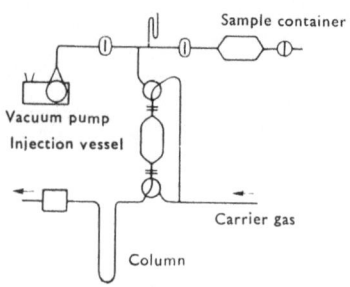

Figure 33. Pressure injection.

Stage 1: evacuate injection vessel, measure vacuum.

2: allow the substance to flow slowly out of the sample container along a capillary into the injection vessel until the desired pressure is reached.

3: by adjusting the Tschako tap or pneumatic valve, pass the carrier gas stream through the injection vessel into the column.

pressure of the gas injected is varied: $p = n \cdot R \cdot T/v$; $p = n \cdot k$. Further details can be obtained from *Figure 33*[6].

Advantages, disadvantages and assessment as for the by-pass system. The cost is much higher (vacuum pump, manometer). The process is, however, worth considering for the injection of samples of a gas for which there are no suitable sealing fluids, and yet which must be taken in a limited volume.

5. *Pneumatically Controlled By-pass Systems*

This process (Hooimeijer, Kwantes and van de Craats[7]) may be regarded as the most elegant of the gas injection processes. It is primarily designed for

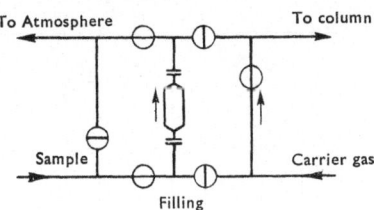

Figure 34. Arrangement of the pneumatically controlled by-pass system of Hooimeijer, Kwantes and de Craats[7].

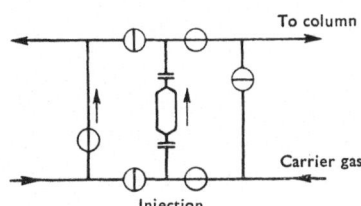

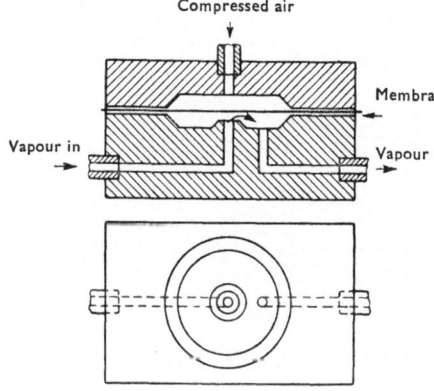

Figure 35a. Cross section through a switch unit of a diaphragm valve to show its method of operation.

automatic instruments, but as it works exceptionally well over long periods without maintenance, it is finding increasing use in simple laboratory instruments. Pneumatically controlled valves of very small volume replace the stopcocks. They require no lubricant. There are no rotating parts. The valves can be controlled satisfactorily with compressed air at 1–2·5 atm gauge pressure. The necessary details are shown in *Figures 34* and *35*a, b and c.

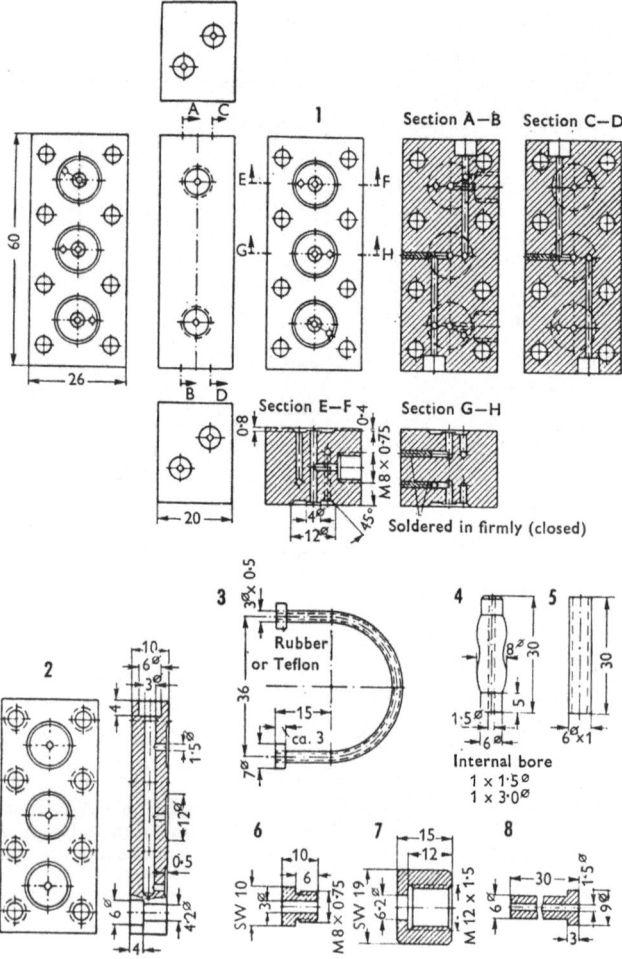

Figure 35b. Recommended construction of the pneumatic sample injector.

Disadvantages. Practically none, if suitable construction materials are used.

Assessment. The somewhat higher cost as compared with the methods described previously is more than compensated for by the outstanding properties of this process. It is the most suitable method for quantitative analysis, and the obvious choice for automatic operation.

The valve bodies are easy to produce. The cavities are sunk into suitable metal blocks with round cutters (using a drill), the middle gas paths having already been drilled. The surfaces of the valve blocks must naturally fit one another well. Cellulose film, silicone rubber or rubber diaphragms from air balloons used for technical purposes (small meteorological balloons) can be used as diaphragm material. To avoid adsorption on the diaphragm surfaces the latter can easily be covered with a substance of opposite polarity.

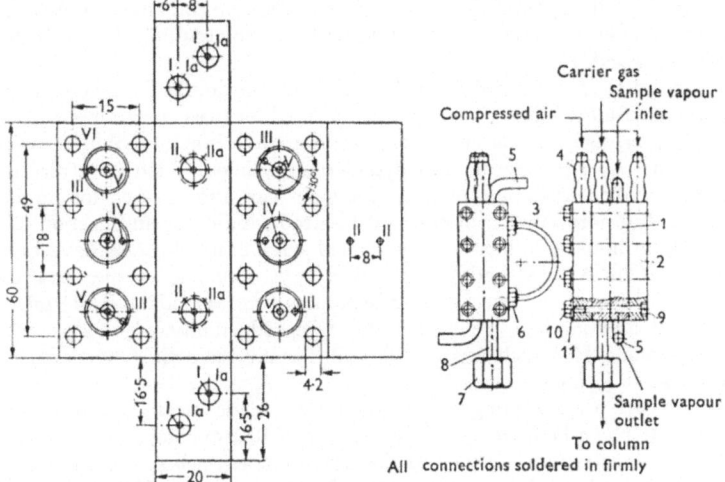

Figure 35c. Drilling plan; assembly diagram for *Figure* 35*b.*

Thus if, for example, hydrocarbons are to be injected, the side of the diaphragm exposed to the substance can be coated with a thin film of polyethylene glycol or painted thinly with 2,2-dimethoxybutanol.

Data for drilling: see *Figure 35c.*

Drill Hole No.	Diameter	Depth	Selection
I	1·5	30	4
Ia	6	4	4
II	1·5	13	4
IIa	M8 × 0·75	6·5	2
III	1·5	14	4
IV	1·5	7	4
V	1·5	Goes right through	4
VI	4·2	Goes right through	8

Another very good system for gas injection is the microdipper, which is described under liquid injection.

For the injection of gases and vapours in quantities of 1γ up to less than $1/1000\gamma$ (10^{-9} g) systems can be used which work on the stream splitting drinciple. See Volume II.

Sample Injectors for Liquids

The injection of liquids is difficult. For one thing, the quantity to be introduced is in the gamma to microgram range, and for another, the losses occurring as a result of premature evaporation or adhesion can alter not only the amount but also to a very considerable extent the composition of the mixture to be analysed.

A number of the systems which have been previously described in the literature are modifications of the systems used for mass spectrometry.

However, such systems are in many cases not a satisfactory solution to the problem (Davis[8]). The medical hypodermic syringes which have already been described as sample injectors for gases can also be used for liquids (Ray[3]), but normally to obtain a good reproducibility for an injection of a few mg of liquid is difficult. Complicated syringes for this task have been described by Adams[9] and Carle[10]; they have very fine needles and are fragile. The dead volume for the needle alone corresponds to 0·004 ml. If 0·015 ml of liquid are to be injected then a reproducibility of $\pm$ 0·2 per cent may be achieved, which is $\pm$ 3×10^{-4} ml. For this there must, of course, be no air at all in the syringe.

With such a microsyringe, which has a very slender steel or teflon piston, an injection can be made against a pressure of 10 atm gauge.

A micrometer screw is used as the piston guide. Such syringes are very expensive (in 1958 they were about DM 500, or approx. £45). In spite of this it is possible to carry out the accurate injection of gamma quantities of liquid without the use of such accurate syringes. Other systems are used, which are described in more detail below.

1. *Micropipettes*

A micropipette (*Figure 36*) is placed under the surface layer of the liquid sample. The capillary effect itself is sufficient to fill it. The gas supply to the column is interrupted, the column head is opened, and the tip of the micropipette is placed on the surface of the packing. The liquid is thus drawn out of the capillary into the packing. The column is closed and the gas stream is turned on again. This method was used by James and Martin.

The smallest amount that can be injected with a micropipette is 0·000025 ml, which corresponds to about 25 gamma. Such pipettes must, however, be emptied with the aid of compressed air.

Advantages. They are very simple. Practically no supplementary materials are required. The glass pipettes can be prepared in the laboratory. Very small amounts of substance can be injected, because the length of the capillary can be kept small. The maximum amount possible is not limited, since injectors resembling ordinary pipettes can be used, although these, of course, have to be filled by suction.

Disadvantages. The introduction of the substance causes a considerable disturbance to the column equilibrium. Volatile components may be lost,

Figure 36. Glass micropipette. Diameters of the capillary openings above and below: 0·7–0·05 mm. Length between 3 and 20 mm.

since the back pressure from the column head may cause gas to flow in the direction of the column head. The injection is not very accurate; the reproducibility is generally over ± 5 per cent. Unequal amounts of adhering liquids are retained in the capillary. Before every sampling the substance under analysis must be carefully homogenized.

2. *Microdippers*

Tenney and Harris[11] have described this process, which works in a very elegant and accurate manner and is recommended. The author has used the microdipper for sample injection at high temperatures; this form will be described. The sample is drawn automatically into a capillary (*Figure 37*). The column head is closed with a special tap.

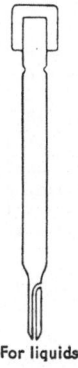

For liquids

Figure 37. Metal microdipper rod. Inserted in the special tap in *Figure 38*. Bore: 0·7–0·2 mm. (Attained where necessary by the insertion of a small piece of fine metal capillary.) Length 3–20 mm.

The filled microdipper rod, which is of metal, is pushed through the gastight rubber ring seal until it is almost at the stopcock. The frictional resistance of the rubber seal is reduced by means of talcum. It seals off the strengthened part of the rod which holds the capillary bore (0·7 mm diameter, 1–18 mm long for 0·5–8 mg liquid).

The stopcock is turned so that the rod can be pushed further in. The lower part of the rod rests finally on the orifice and seat in the column head, thus

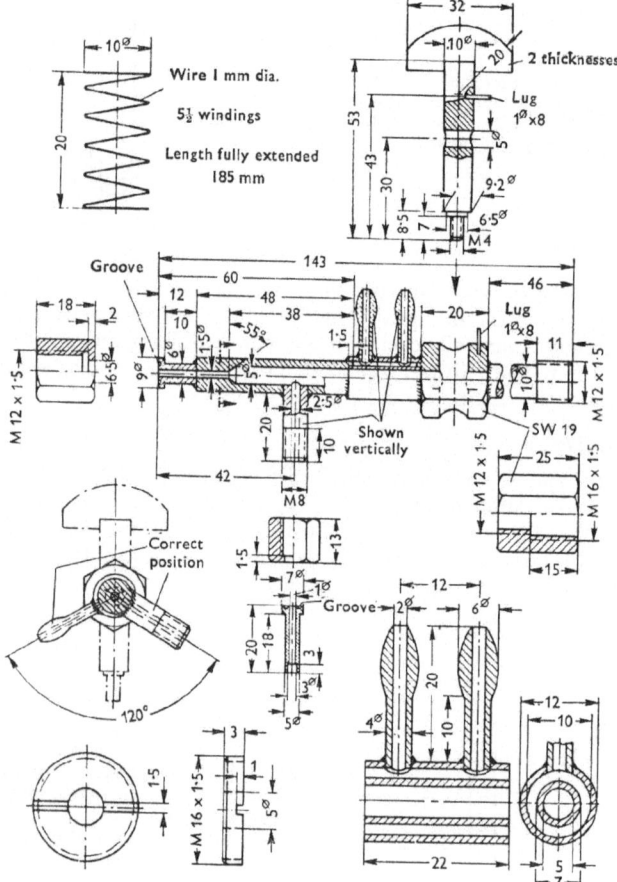

Figure 38. Recommended construction for the microdipper tap.

blocking the passage of the carrier gas into the column head. The pressure increases, the carrier gas drives the liquid out of the capillary, which has in the meantime been heated by the so-called flash heater so that all the liquid adhering to the capillary walls is vaporized and washed out into the column.

The rod is now withdrawn until it is just above the stopcock, the tap is closed and the rod is removed completely from the guide and the ring seal. The 'depth of immersion' necessary to reach the stopcock may easily be recognized by making a mark on the rod. The temperature of the stopcock is con-

24 V, 25W

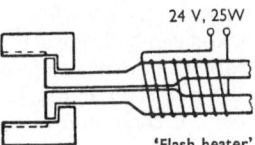

Figure 39. Flash heater for the temperature control of the point of evaporation in the sample injector.

'Flash heater'

trolled by a water cooling system, a capillary in the water supply enabling the maintenance of a constant and correct amount of cooling water of *ca.* 3 drops/sec. Because of this the microdipper can also be used for the injection of high-boiling solids, where the column and flash temperatures are quite high and the metal tap could also, by thermal conduction, become very hot. Solids (solid paraffins, long-chain acids, alcohols, phenols, etc.) are melted,

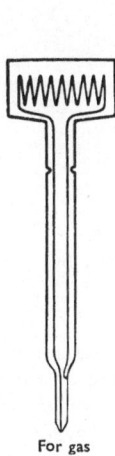

For gas

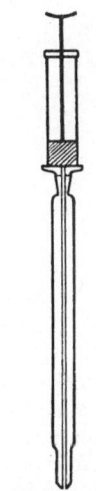

Figure 40. Microdipper for gas. The gas spiral can be made of tubing with 0·3 to 3 mm internal diameter, with 1 mm i.d. as the most suitable.

Figure 41. Microdipper for larger amounts of gas or liquid, especially where such amounts are variable. The bore (long injection canula in stronger tube) is not washed out.

the capillary is warmed, and the substance is drawn up into it; the process then continues as described above. It does not matter if in the meantime the substance resolidifies.

Advantages. The preparation of the microdipper is not difficult. It may be made either of glass or of metal (steel stopcock in a brass guide, or the like). The reproducibility of quantitative injections lies in the region of ± 0·5–0·8 per cent. It can be increased to ± 0·3 per cent. It is just as suitable for solids and liquids as for gases. For the latter application the form of the rod is slightly modified (see *Figure 40*).

Since the rod can be made of glass the microdipper can be used for corrosive materials. The quantity injected can be increased from microquantities to grams if the variant shown in *Figure 41* is used. The substance under analysis does not come into contact with any lubricant (for the stopcock

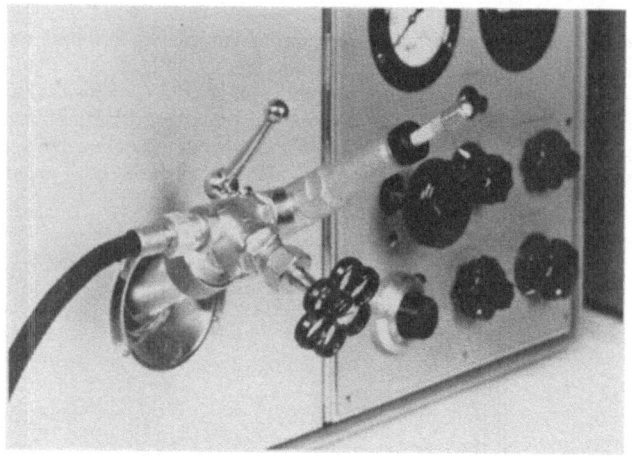

Figure 42. Sample injector for liquids manufactured by Perkin Elmer GmbH Bodenseewerk.

etc.) which can also be of importance when the method is used for gas injection, the point of injection can be as hot as desired and the sample can solidify in the capillary without causing any trouble. When introduced into the hot stem of the injection tap it will melt again and evaporate.

The microdipper can be used at practically any temperature and can be applied to pressurized and vacuum gas chromatography, although in this case the grinding of the stopcock must be very accurate. A commercial variant is shown in *Figure 42*.

Disadvantages. The process cannot be made automatic. The stopcock requires careful attention and precise design. The use of the method requires some skill (if the tip of the rod touches the stopcock before the latter is opened the liquid will run out of the capillary, dilute the stopcock lubricant, and enter the column slowly and irregularly). Substances which have both abnormal surface tensions and high specific weights can be injected only imperfectly, if at all.

3. *Pneumatically Controlled By-pass Systems*

This process, which has already been described in detail on p. 87 and in *Figures 34–35c*, was suggested by its authors as also capable of application to liquid injection. The method is, in particular, exceptionally good for the injection of liquids with high vapour pressures, aggressive liquids, and liquids which cannot be allowed to come into contact with air, and it may also be regarded as the only process available at present which can be made automatic. The apparatus is combined with a vaporizer which converts the liquid into vapour. The vapour stream is injected in the same way as a gas, the temperature of the apparatus being kept above the dew point. The membrane material available at present limits this variant to operating temperatures of 180–250°C. Substances with higher boiling points can be injected as

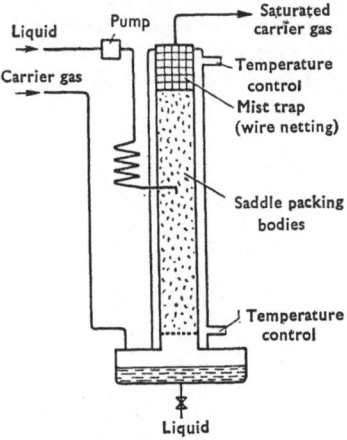

Figure 43. Saturation 'tower' for the injection of high boiling liquids by the automatic method with the pneumatic sample injector shown in *Figures 34* and *35.*

follows. A continuous stream of liquid flows dropwise through a 'saturating column'. (See *Figure 43.*) This column, which is 10–20 cm high and 2 cm wide, is filled with suitable packing material (e.g. saddle-shaped packing bodies as used in distillation columns) and heated to a given suitable temperature. A slow constant gas stream (branch) flows countercurrent through the column and becomes saturated with the vapour of the liquid in a manner that corresponds to the composition of the liquid phase, or to the composition of the mixture of vapours resulting from it, the temperature of the saturation column determining the concentration. This vapour is injected in the same way as a gas.

In the case of very high boiling liquids the composition of the vapour no longer corresponds to the composition of the liquid phase. However, in continuous process control an analysis of this kind, which can be carried out completely automatically, can still be of use, since the thing of interest here is generally any variations in the process, which can still be recognized with certainty.

Advantages. The process can be made automatic. It is especially suitable for liquids with high vapour pressures, and it is also suitable for volatile liquids which are otherwise difficult to inject.

Disadvantages. The application is limited to materials with boiling points below 250°C. If suitable membrane materials (beryllium-copper etc.) are used the range of application may be extended. The saturating tower requires very careful temperature control, and the gas flow rate to the tower must be carefully balanced with the liquid flow rate. The higher boiling fractions are no longer taken up in a manner directly proportional to the true composition.

4. Capillary Crushers

This process, which was described by McCreadie and Williams[12], is especially suitable for cases where it is absolutely essential that the quantity of liquid injected should be exactly known, or where very small quantities of vapour are to be injected. The substance is drawn into a capillary; this is closed at both ends and put into the so-called capillary crusher (see *Figure*

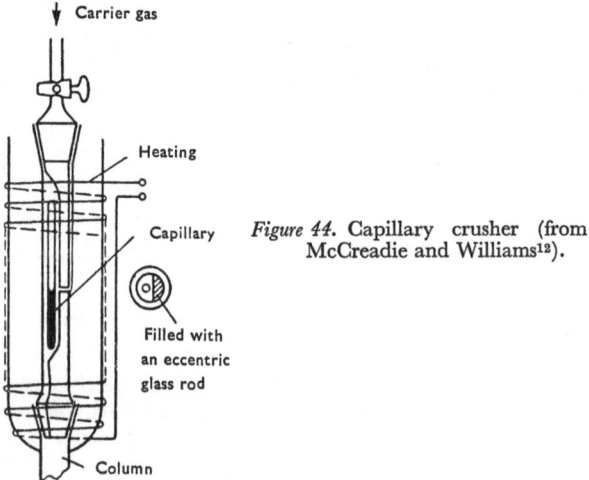

Figure 44. Capillary crusher (from McCreadie and Williams[12]).

44). The gas stream, which has been cut off, is now switched on again, and left until complete equilibrium is regained. The upper part of the apparatus is now rotated, which crushes the capillary by grinding it against the stationary lower slit of the apparatus. The capillary crusher can be heated, so that the substance evaporates as soon as it is released. The dead volume can be kept very small.

Advantages. Optimum behaviour according to the theory of sample injection, with quantitative data whose precision depends only on the precision of the microbalance available. Suitable for substances which must be protected from air or moisture.

Disadvantages. Sample injection is a very tedious process and requires supplementary materials: balance, capillaries that need to be sealed. It is not possible to carry out comparative measurements, because for these the amount of substance injected must always be identical. Low-boiling substances can only be sealed into small capillaries with difficulty. The danger of variation in composition is great. For heat-sensitive materials the process is practically out of the question.

Piston Injector

The author uses the piston injection system shown in *Figure 45* for the automatic injection of medium to high boiling substances flowing continuously through a pipeline.

Figure 45. Piston injector for liquids.

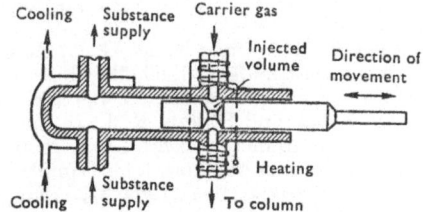

Advantages. It answers the need for an automatic injection process for liquids of medium to high boiling point.

Disadvantages. The substances which can be injected are limited to those for which there is a resistant non-swelling sealing material. The amount injected is not variable, or at least, is troublesome to alter.

References

1. PORTER, P. E., DEAL, C. H. and STROSS, F. H., *J. Amer. Chem. Soc.*, 1956, **78**, 2999.
2. HARRISON, G. F., *Vapour Phase Chromatography*, ed. D. H. Desty, Butterworths, London, 1957, p. 332.
3. RAY, N. H., *J. Appl. Chem.*, 1954, **4**, 21.
4. KEULEMANS, A. I. M., *Gas Chromatography*, Reinhold Publ. Corp., New York, 1957, p. 66.
5. Private communication from Dr. Kögler, Böhlen.
6. Private communication, 1st International Symposium on Gas Chromatography, London 1956.
7. HOOIMEIJER, J., KWANTES, A. and CRAATS, F. VAN DE, *Gas Chromatography*, ed. D. H. Desty, Butterworths, London, 1958, p. 288.
8. DAVIS, R. E. and McCREA, J. M., *Analyt. Chem.*, 1957, **29**, 1114.
9. ADAMS, N. G., Prospectus, Ethyl Corp., Baton Rouge.

10. CARLE, D. W., *Gas Chromatography*, ed. Coates, Noebels, Fagerson, Academic Press Inc., New York, 1958, p. 67.
11. TENNEY, H. M. and HARRIS, R. J., *Analyt. Chem.*, 1957, **29**, 317,
12. McCREADIE, S. W. S. and WILLIAMS, A. F., *J. Appl. Chem.*, 1957 **7**, 47.

Further literature

BRADFORD, B. W., HARVEY, D. and CHALKLEY, D. E., *J. Inst. Petrol.*, 1955, **41**, 80.
CRAATS, F. VAN DE, *Analyt. chim. acta*, 1956, **14**, 136.
DIMBAT, M., PORTER, P. E., STROSS, F. H., *Analyt. Chem.*, 1956, **28**, 290.
DUBSKY, H. E., JANAK, J., J. *Chromatog.*, 1960, **4**, 1, and a further 41 literature references.
FREDERICKS, E. M. and BROOKS, F. R., *Analyt. Chem.*, 1956, **28**, 297.
GLEW, D. N. and YOUNG, D. M., Stopcock for gas chromatography, *Analyt. Chem.*, 1958, **30**, 1890.
JOKLIK, J. and BAŽANT, V., Capillary tube crusher for use in gas chromatography, *Chem. Listy*, 1959, **53**, 277.
LICHTENFELS, D. H., FLECK, S. A., and BUROW, F. H., *Analyt. Chem.*, 1955, **27**, 1510.
MALMSTADT, H. V. and HICKS, G. P., Rapid injection and automatic refill pipet, *Analyt. Chem.*, 1960, **32**, 445.
POLLARD, F. H. and HARDY, C. J., *Chem. & Ind.*, 1955, 1145.
PURDY, K. M. and HARRIS, R. J., Introduction of liquid samples into the mass spectrometer, *Analyt. Chem.*, 1950, **22**, 1337.
SAMSEL, E. P. and ALDRICH, J. C., Sample injection valve for gas chromatography, *Analyt. Chem.*, 1959, **31**, 1288.
SWEETING, J. W., Modification of Agla micrometer hypodermic syringe for use in vapour-phase chromatography. *Chem. & Ind.*, 1959, 1150.
STANFORD, F. G., Sample-injection method for gas-liquid chromatography, *Analyst*, 1959, **84**, 321.

Further literature references on sample injectors will be found in *Gas Chromatography Abstracts* for 1958, 1959, 1960, and following volumes, under the subject index number 3.2.

2.4. DETECTORS

The following section deals with the necessary and desirable features of detectors, gives short descriptions of the various principles of measurement, and provides extensive data on detectors which may be easily constructed.

In conclusion there is a description of methods of testing detectors in order to be able to give a quantitative description of their suitability in practice.

DEFINITION: the term detector is understood to mean an instrument that converts the chromatographic result into a form in which it can be recorded. In combination with the recorder the detector provides the so-called gas chromatogram.

One of the reasons that gas chromatography has spread so rapidly is that it enables a substance in a gas stream to be detected easily and with an astounding sensitivity. In almost all cases the signals obtained by the detector may be converted into electrical signals. The analytical result is easy to record, and has the value of giving a permanent record.

It is useful to distinguish two types of detectors: the first kind measure the substance directly, while the second give a measure of the combined properties of the carrier gas and substance. In any case it will be realized that the separated substances leaving the column are mixed with a large excess of carrier gas.

Detectors of the first type are basically to be preferred to those of the second type, since the measurement obtained refers to the substance alone and not to a mixture of substance and carrier gas.

It is also helpful to classify them into two other groups, depending upon the type of result recorded. If the detector records the concentration of the given substance as it varies from moment to moment, i.e. the differential of the analytical result, it is called a differential detector. Together with the recorder it produces the differential chromatogram, which consists of peaks.

If, however, the detector continually adds together the amounts of substances issuing from the column it is an integral detector and the result is an integral chromatogram, consisting of a step-shaped curve.

In both cases the result recorded may be evaluated quantitatively. For differential chromatograms the area under the peaks corresponds to the amounts of the appropriate substances present; for integral chromatograms the step height is a measure of the amounts of the corresponding substances.

The degree of precision with which the areas thus obtained correspond to the true concentration depends among other things on the construction of the detector and its principle of measurement.

Desirable and Necessary Properties of Detectors

The result of the chromatographic separation in the column must be accurately monitored by the detector and should be truly recorded by the recorder. As we can only judge the quality of the result from the recorded curve we shall in the following section regard the detector and recorder as one unit and assume that the recorder behaves perfectly. It can thus not give rise to any additional errors. From the above requirement that the result should be recorded accurately, the following necessary and desirable properties for a detector may be derived:

1. The detector must reveal what is occurring in the carrier gas stream from moment to moment without any interference from a time lag, i.e. with a delay of less than $\sim$ 1 sec. (The best time at present is 8 milliseconds with the flame ionization detector.)

2. The volume of the detector must be so small that the substances separated in the column have no chance to remix. It is expedient to have it about the size of the volume of substance vapour that can just be taken up by one

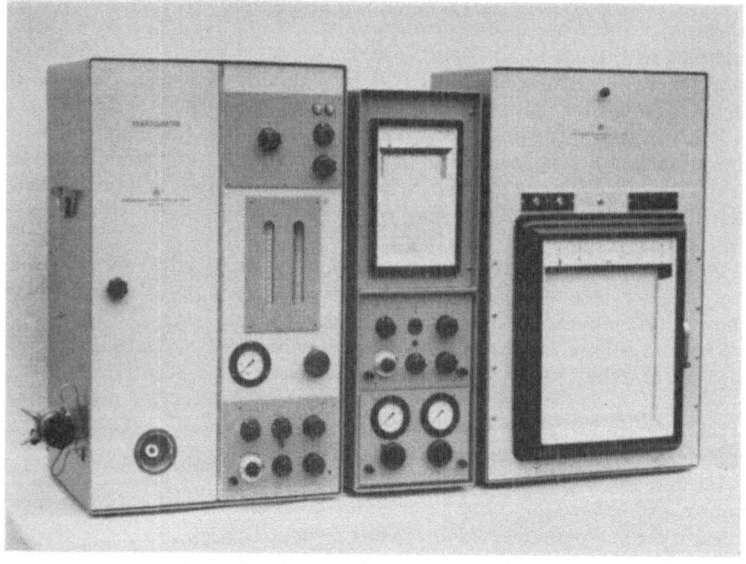

Figure 46. Gas chromatograph manufactured by Perkin Elmer GmbH, Boden-seewerk, on the unit construction principle, which enables the simultaneous use of two detectors (e.g. thermal conductivity cell and flame ionization detector).

theoretical plate. This quantity corresponds to what was called in Section 2.1 the maximum load capacity of the column. For detectors connected to packed columns the measuring volume should therefore be 0·1–0·5 ml. (The best value at present is much less than 1 microlitre.)

3. Values independent of the substance, such as pressure, temperature, and gas flow rate, should not have a great influence on the signal.

4. The dependence of the detector signal on the concentration should within wide limits be linear, i.e. the detector should have a linear signal when the concentration of substance in the gas stream varies between 0 and 1 per cent.

5. The detector should have a high sensitivity, and should be able to show the presence of 10^{-6} g substance in 1 ml of carrier gas with certainty. (The best value at present is 10^{-14} g/ml.)

There are different detectors for different methods of operation and different types of separation, and these factors should be borne in mind when selecting the detector.

There is no ideal detector which can fulfil all the requirements equally well.

Besides the task of 'converting' the chemical substances into electrical values for recording purposes (without necessarily destroying the substances) the detector can also perform a valuable service in the determination of chemical constitution. In particular, when a non-specific detector is connected in series with a second, specific, detector so that the same substance passes through each detector in turn, very valuable information may be obtained. An example of a commercial instrument which takes into consideration this possibility of working with two detectors is shown in *Figure 46*.

Principles of Measurement

All the physical, chemical (and at present even biological) properties of a two-component system, consisting of a large excess of gas and a small

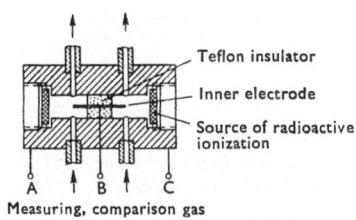

- Teflon insulator
- Inner electrode
- Source of radioactive ionization

A ↑ B ↑ C
Measuring, comparison gas

Figure 47. Principle of the β-ray ionization detector.

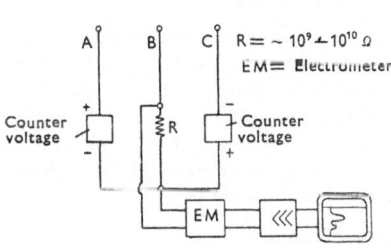

amount of substance dispersed in that gas, have been used as the fundamentals of principles of measurement. In the following, only those principles which are used in practice are described; for more detailed information the specialist literature should be consulted.

Special attention is paid to the thermal conductivity cell, and practical data are given on the construction of a very high sensitivity flame ionization detector.

1. β-ray Ionization

An organic compound may be directly or indirectly ionized by means of high energy radiation (e.g. β-rays). In a field of electrical potential the ions produce an ionization current which can be amplified and measured. The ionization can be carried out indirectly by means of excited high energy inert gas atoms. (Deal *et al.*[1], Boer[2], Lovelock[3]) (*Figure 47*). For further details see Volume II.

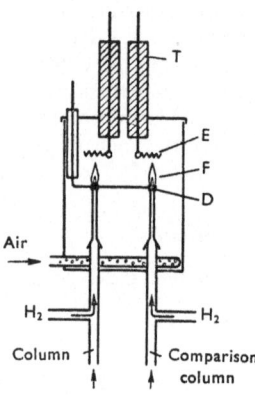

Figure 48. Two-flame ionization detector (from McWilliam and Dewar[4]), now used with only one flame.

2. Flame Ionization

By burning the substance in a hot flame ions are formed, which in a potential field form an ionic current which can be amplified and recorded (McWilliam[4]) (*Figure 48*). For further details see p. 112.

3. Wireless Valve Detector

If the ionization potential of the carrier gas is sufficiently different from that of the organic substance then opened wireless valves can be used as detectors. The valve (e.g. RCA type 1949) has a substance inlet, through which passes about 0·5 per cent of the gas stream from the column. By means of an attached vacuum pump the pressure in the valve is kept to 0·2 mm Hg. There is a filament emission of 5 mA (iridium cathode), a grid potential of + 18 V and an anode potential of + 27 V (Ryce and Bryce[6]).

4. Flame Emission

If the vapour of the substance is burned, colours will appear in the otherwise practically colourless hydrogen flame. The intensities of these colours can be recorded by photocells (Grant[5]) (*Figure 49*).

Such a detector has a small volume. It is admittedly not very sensitive, but can be particularly useful for preparative instruments[24].

5. *Optical Absorption*

The absorption spectrum in the infra-red to ultra-violet wavelengths is similar for many organic molecules at a number of wavelengths. Detectors consisting of a suitable light source and a suitable receiver (e.g. infra-red

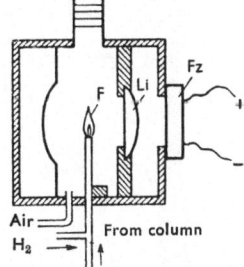

Figure 49. Flame emission detector (Grant[5]).

absorption spectrograph) may be used. Besides measuring the absorption of the substance itself, the absorption of the CO_2 produced by burning the substance over CuO may also be measured (*Figure 50*).

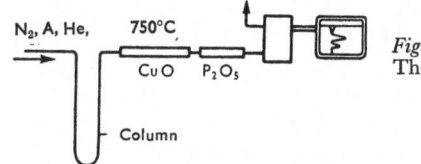

Figure 50. CO_2 — URAS as detector. The burned substance is recorded as CO_2.

Infra-red detectors work satisfactorily at a wavelength of 3,330 cm^{-1}, which gives the least possible distortion in the C-H band [23].

6. *Chemical Reactions, Solubility, Heat of Solution*

Whether CO_2 is used as the carrier gas and the fact that the organic gases are not absorbed in caustic potash is used as the measurement principle, or whether the substance is converted by combustion into CO_2 and this is measured by titration (e.g. coulometry or the change of colour of an indicator solution) depends on the task itself. On this basis detectors of high sensitivity can be built (*Figures 51* and *52*).

However, all these types of detector are inferior to those given under 1, 2 and 3 as regards sensitivity.

The heat of absorption (heat of solution) of the substances passing through the column can also be used as a measurable quantity. Thus a very simple detector can be made from a sensitive thermo-electric pile which is placed at the column outlet within the column packing. The second junction of the thermopile is placed on the outer wall of the main column. The voltages can be recorded by a continuous line recorder in the usual way.

7. *Thermal Conductivity*

The thermal conductivity of hydrogen or helium is about 6–10 times as great as that of any organic vapour. The thermal conductivities of other

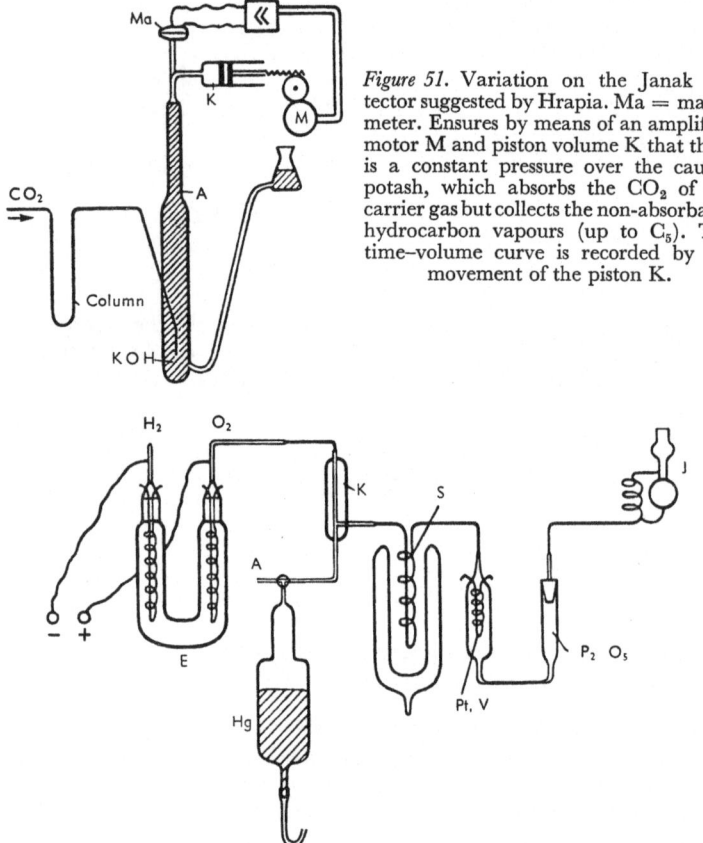

Figure 51. Variation on the Janak detector suggested by Hrapia. Ma = manometer. Ensures by means of an amplifier, motor M and piston volume K that there is a constant pressure over the caustic potash, which absorbs the CO_2 of the carrier gas but collects the non-absorbable hydrocarbon vapours (up to C_5). The time–volume curve is recorded by the movement of the piston K.

Figure 52. Juranek detector for traces of hydrocarbons in air. Carrier gas O_2; E = O_2 gas developer; A = sample intake point; K = catalyst for decomposition of ozone; S = column; Pt, V = glowing platinum wire in combustion chamber; I = CO_2 absorption solution coloured with phenolphthalein, the colour change of which is recorded by means of photo-electric cells.

suitable gases are either greater or lesser than those of the other organic vapours.

It is only the relative fluctuations in the thermal conductivity of gas mixtures that can be easily measured. For this the relative temperature fluctuation of a very thin, electrically heated wire or a very small (microscopic point) resistance surrounded by the carrier gas-substance mixture is measured. Let us assume that the carrier gas issuing from the column is pure hydrogen. This substance has a very high thermal conductivity. The wire in the thermal conductivity cell is therefore strongly cooled. Now suppose that the carrier gas is mixed with benzene vapour. The mixture now has a considerably lower thermal conductivity than the pure carrier gas. The wire is less strongly cooled than before, i.e. it becomes warmer. But at the same time its electrical resistance is also altered, which can, by means of an appropriate circuit, easily be converted into a variation in voltage. Further details on thermal conductivity are given on p. 107.

8. Heat of Combustion

If the vapour of the substance is burned in a hydrogen flame both the flame temperature and the flame size are increased. This will alter the temperature of a thermocouple junction placed over the flame, and this change in temperature will be converted into a change in voltage (Scott[7]) (see *Figure 53*).

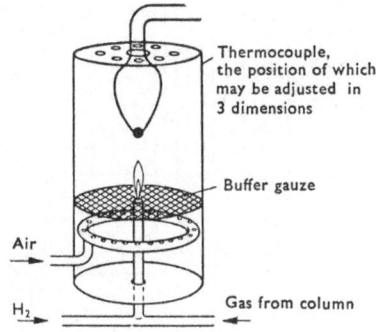

Thermocouple, the position of which may be adjusted in 3 dimensions

Figure 53. Micro flame detector (Scott[7]).

Buffer gauze

Air

H₂

Gas from column

9. Velocity of Sound

The great differences in the velocity of sound in different gases enable the detection of, for example, 0·0006 per cent of hydrogen in air and 0·005 per cent of oxygen in nitrogen. Sound waves are produced at one end of the measuring cell and received by a crystal at the other end. The phase displacement of the wave received with respect to the wave transmitted gives an exact measure of the alteration in the velocity of sound which depends on the composition of the gas[8].

10. Dielectric Constant, Viscosity, Surface Potential on Condenser Plates

These three methods were proposed by Griffiths, James and Phillips[9] as detector principles. As, however, they do not have special advantages, and high sensitivity detectors cannot be constructed by means of them, a mere mention here will suffice.

11. *Gas Density*

The measurement of gas density was successfully used as the basis for a detector by Martin with his famous Gas Density Balance, a highly sensitive instrument which can be used without specific calibration. The reference gas and the gas under investigation flow through four channels. These channels are linked in a manner similar to that of a Wheatstone bridge. If differences of density occur in the two gas streams they are compensated for by a cross stream. This causes heated gas to pass over a thermocouple, so that the temperature variations are converted into voltage variations (Martin and James[10]) (see *Figure 54*).

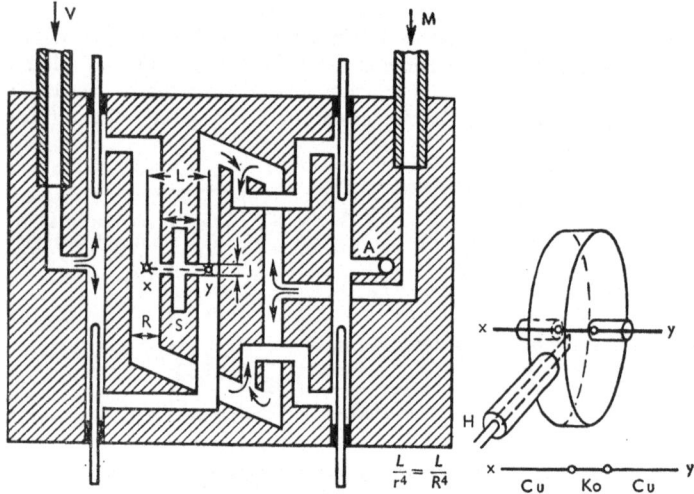

Figure 54. Martin gas density balance (Keulemans, *Gas Chromatography*, Reinhold Publ. Corp., New York, 1957).

V = comparison gas inlet   S = measuring element
M = measuring gas inlet   Cu—Ko—Cu = thermoelement
A = common gas outlet   H = heating element
The position of the measuring element is shown separately (x — y).

Griffin and George Ltd.(Ealing Road, Alperton, Wembley, Middlesex) are now producing these excellent instruments on an industrial scale, which will permit wider application of an instrument which is superior to the thermal conductivity cell. For a vapour with the molecular weight M in a carrier gas with a density of q_0 (g/l.), the difference in density is equal to

$$q - q_0 = C \cdot \frac{1 - 22 \cdot 4 \ (q_0/M)}{1 + 22 \cdot 4 \ (C/M)}$$

C is the measured concentration of the vapour in g/l. of the carrier gas measured under normal conditions. As in most cases the factor C/M is very small, the density difference $q - q_0$ is directly proportional to the weight concentration.

The detector can be used for concentrations of as little as 0·1 g/l. without an amplifier and down to $1·5 \times 10^{-5}$ g/l. with one.

For the analysis of CO_2 this corresponds to a lower concentration of 20 p.p.m. at full scale deflection, or to 15 p.p.m. at full scale deflection in the case of ozone in air. The efficiency of this detector has, up to now, not received the attention it deserves, because of the constructional difficulties.

12. *Electron Capture Detector*

This detector was first described by Lovelock and Lipsky[31], and has the special property of being substance specific. The signals produced are dependent on the type of functional groups in the molecule. In an ionization chamber, which contains ions produced by radioactive irradiation, the decrease in ions due to ion capture, which is specifically influenced by the functional groups, is measured. Functional groups containing atoms other than carbon and hydrogen are particularly easy to distinguish. Interesting detection possibilities are created by combining this detector with a non-selective detector. This detector is almost as sensitive as a flame ionization detector.

The Thermal Conductivity Cell

Schleiermacher was probably the first to carry out the measurement of the thermal conductivity of gases and vapours by means of the so-called hot wire method.

Although even at the Second International Congress on Gas Chromatography in Amsterdam in 1958 Keulemans described the thermal con-

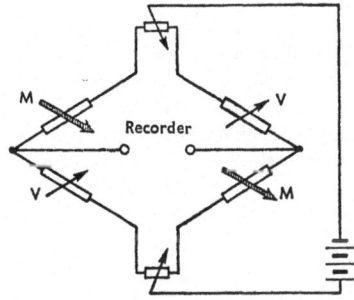

Figure 55. 'Crossover' arrangement for the measuring elements in an electrically symmetrical thermal conductivity cell,

M = measuring gas
V = comparison gas

ductivity cell as a detector of the past, it is likely to find a great many applications in the future in simple control and measurement instruments, especially in automatic instruments, because of its robust and simple construction and because of the other advantages given below.

Carefully constructed cells are largely insensitive towards fluctuations in temperature, pressure and flow rate. This is achieved by constructing the cell so that it is completely symmetrical from the mechanical, electrical and thermal viewpoints. By choosing the correct gas path arrangement one can also help to ensure a good stability from the point of view of application. The arrangement shown in *Figure 55* follows the above principle, and also, simply by means of the arrangement, gives the greatest sensitivity.

Although this fact is well-known there are cells produced even today, by reputable firms, which are largely unsymmetrical in their construction or arrangement and thus are highly sensitive towards fluctuations in tempera-ture, gas flow rate, and the other operating conditions. Even if such errors do not show themselves as variations in the base line they will occur as variations in sensitivity and are thus much more insidious, since they cannot be detected unless a very careful check is made.

Unsymmetrical cells also have no linear signal in the first place, which in particular affects quantitative analyses, unless very small quantities of sub-stance are used; but thermal conductivity cells always require more analy-tical substance than ionization detectors.

Theoretical Considerations

A thermal conductivity cell always consists of a hot central area and a cold surrounding wall. Heat transfer therefore takes place from the hot central area outwards, and this is the more intensive, the greater the tem-perature difference and the greater the thermal conductivity of the carrier gas. The effect measured is created by the change in temperature at the hot central area. If the gas flowing through the cell has a high thermal conduc-tivity then the central area cools rapidly. If the thermal conductivity of the gas alters then the temperature of the hot central area also alters. In order to make as much heat energy as possible available for this process, heat losses due to conduction by the cell material or to radiation should be kept rela-tively low. The cell is the more sensitive, the smaller the quantity of sub-stance that is necessary to cause a large temperature change at the hot central area. As the thermal conductivity of the vapour is fixed, the large temperature change required can only be achieved by keeping the mass and the specific heat of the hot central area very small, although for an intensive exchange the surface must be very large. It is therefore an advantage to use not a wire but a ribbon, e.g. of dimensions $4 \times 45 \mu$. A platinum-nickel or platinum-iridium alloy (95/5) is a suitable material. Ribbons are mechani-cally more stable than wires.

It would not appear necessary to give further constructional details, as there is a wide variety of different types of cells available. (Otherwise see Kaiser, *Gas-Chromatographie*, Akademische Verlagsgesellschaft Geest & Por-tig KG, Leipzig, 1960, pp. 103–111.)

Even in recent years a number of papers have been published on problems connected with thermal conductivity cells. This shows that the evaluation of the results and the correct application of such cells is not a simple matter.

The Thermal Conductivity Cell

Schmauch and Dinerstein[30] compared cells using hot wires and cells using thermistors as measuring element.

The authors give two factors for the sensitivity of thermal conductivity detectors:

1. the cell factor C_w (wire cell) or C_T (thermistor cell),
2. the thermal conductivity factor.

The following relationships hold good for the factors:

for the hot wire cell:

$$C_w = \sqrt{\left[\frac{2\pi L}{\ln(r_W/r_R)} \cdot \frac{I}{R_M}\right]} \cdot \alpha \cdot \left(\Theta_R - \Theta_W\right)^{3/2} \cdot \frac{R_0 \cdot R_1}{R_1 + R_M}$$

for the thermistor cell:

$$C_T = \sqrt{\left[\frac{2\pi L}{\ln(r_W/r_R)} \cdot \frac{I}{R_M}\right]} \cdot \left(\frac{-B \cdot R_M}{R_0 \cdot T^2}\right)\left(\Theta_R - \Theta_W\right)^{3/2} \frac{R_0 \cdot R_1}{R_1 + R_M}$$

where

L	= length of the measuring element
r_W, r_R	= radius of the measuring chamber or element
I	= Joule's equivalent of heat
R_M, R_0, R_1	= Resistances of the measuring elements at $T°C$, $0°C$ and the comparison resistance in the bridge
$\Theta_R - \Theta_W$	= temperature difference between the hot central area and the wall
α	= temperature resistance coefficient
B	= constant for the thermistor material
T	= temperature in $°K$

The relationship between the voltage values of the detector and the thermal conductivities is given by the following equation:

hot wire cell:

$$E_0 \simeq C_w \left[\sqrt{(K_2)} \cdot \frac{K_2 - K_M}{K_M}\right]$$

thermistor cell:

$$E_0 = C_T \left[\sqrt{(K_2)} \cdot \frac{K_2 - K_M}{K_M}\right]$$

where

K_2	= thermal conductivity of the carrier gas
K_M	= thermal conductivity of the substance/carrier gas mixture:

$$K_M = \frac{K_1}{1 + A_{12}\dfrac{1-x}{x}} + \frac{K_2}{1 + A_{21}\dfrac{x}{1-x}}$$

where

K_1	= thermal conductivity of the vapour of the substance
x	= mole fraction of the substance
A_{12} or A_{21}	= Wassiljewa constants

109

As the temperature of the wire at constant wall temperature not only alters the sensitivity of the cell as it increases (and almost invariably alters it by increasing it), but also can cause a considerable alteration in the thermal conductivity of the substance/carrier gas mixture, the deflection of the thermal conductivity cell is not always linearly dependent on the concentration of the substance in the carrier gas. Because of this, at critical values in carrier gases of low thermal conductivity the phenomenon of peak inversion may occur, which can even suppress the detection sensitivity for a given substance or a given concentration.

Because of this, thermal conductivity can only be used without special control measures when helium or hydrogen is used as carrier gas, and even then for quantitative analysis all the apparatus parameters must be kept to a high degree of constancy.

When rare gases are used it must be remembered that the thermal conductivity of monatomic gases increases less rapidly with increase in temperature than, e.g. the thermal conductivity of organic vapours. Thus the increase in sensitivity with increasing wire temperature is very rapid in argon but only slow in helium. It is, however, the relative variations which must always be considered.

Values for the thermal conductivity of different materials are given in Volume III (Tables), together with correction factors for quantitative evaluation. Thermal conductivity cells which only use one resistance of the Wheatstone bridge network as measuring element always give a non-linear

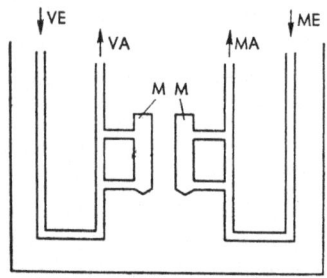

Figure 56. Plan view of a mechanically symmetrical thermal conductivity cell.

VE, VA = comparison gas inlet and outlet.

ME, MA = measuring gas inlet and outlet.

M = measuring cell, split stream form.

signal; they have only a small dynamic range. There is no fundamental difference between the thermistor cell and the wire cell. Wire cells can, however, be used over a wider range (of temperatures as well as concentrations) than thermistor cells, and thermistor cells have a steep sensitivity maximum that depends on the external temperature and the heating current. Generally speaking, they are inferior to hot wire cells.

Influence of the form of the gas path in the cell on the response

Cells may be divided into flow-through, convection-diffusion, and self-purging cells.

The flow-through cell has the advantage that the signal is obtained very quickly, but the disadvantage that the cell measures at the same time variations in gas flow rate. The flow-through cell is not suitable for quantitative analytical work unless a completely constant gas flow rate can be guaranteed.

The convection-diffusion cell is not sensitive towards variations in flow rate, but its signal is hardly linear towards fairly large variations in concentration, so that this type again is not suitable for quantitative analytical work.

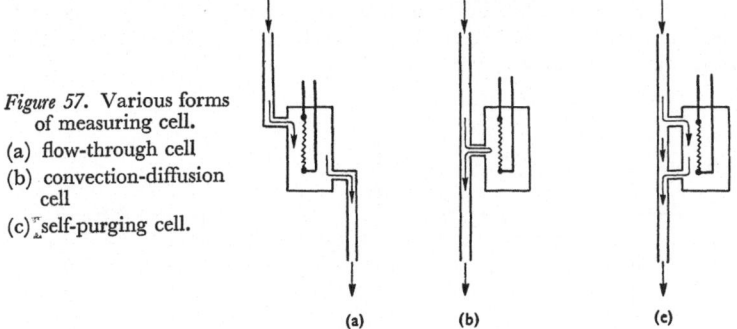

Figure 57. Various forms of measuring cell.
(a) flow-through cell
(b) convection-diffusion cell
(c) self-purging cell.

(a)   (b)   (c)

The self-purging cell is a compromise between the flow-through and the convection-diffusion cell; but there is a wide scope for variations in the arrangement and dimensions of the branched flow paths.

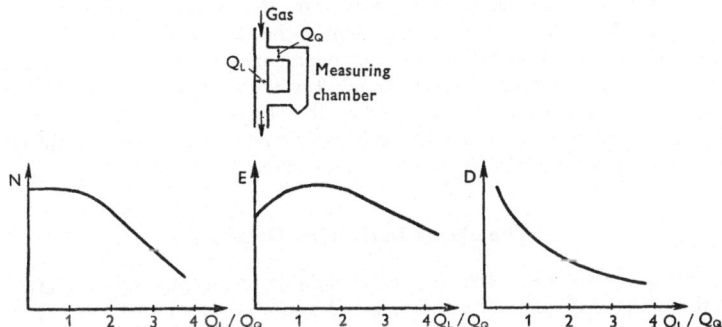

Figure 58. Relationship between the geometric ratios at the split stream point (Q_L= cross section of longitudinal bore; Q_Q = cross section of transverse bore) and the effect on the number of theoretical plates N of the column, the sensitivity E of the detector, and the average quantity D (per cent) of measuring gas through the measuring cell itself.

The author has investigated the conditions and found that the relationships between the sensitivity, the effective cell volume (which affects the column performance) on the one hand, and the ratio of the cross sections of the main and subsidiary gas paths on the other are as shown in *Figure 58*.

The self-purging cell is thus to be preferred to the two other types of cell.

Although the volume of a thermal conductivity cell can be kept very small if thermistors are used instead of wires, there is no point in using thermistor cells except at lower temperatures. The characteristics of thermistors rule out a high operating temperature. Thermistor cells have a strong tendency to give non-linear signals at higher concentrations of substance, but they are not so definitely sensitive to gas flow rate as wire cells. For this reason it is not worthwhile to increase the sensitivity of wire cells with amplifiers, since the noise (caused by heat eddies around the relatively long wire) is much too great.

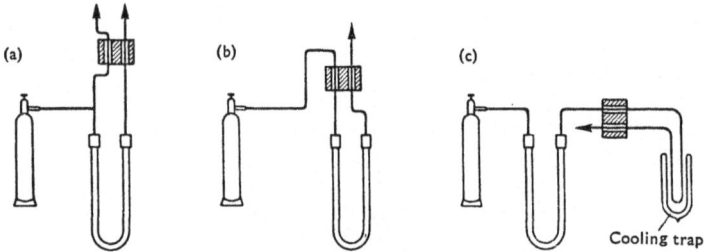

Figure 59. Gas path arrangements for the thermal conductivity cell or for other closed detectors.

The high pressure and flow sensitivity of the thermal conductivity cell can be influenced by the correct arrangement of the carrier gas flow outside the cell, as shown in *Figure 59*. The worst arrangement is that shown in *Figure 59*b, although it is very widely used; *Figure 59*a gives a better solution, but the best of all (although more intricate than the others) is that shown in *Figure 59*c, provided that the analytical substance can be easily and completely frozen out (Dimbat, Porter, Stross[15], Keulemans[16]).

The Flame Ionization Detector

One of the most sensitive detectors available at the present time (1963) is the flame ionization detector. Under favourable conditions it is capable of measuring 0·001 p.p.m. direct, i.e. concentrations of 10^{-10} per cent of impurities in a directly measurable substance (Andreatch and Feinland[26]).

A highly sensitive and very robust detector which can be used at temperatures up to 400°C can be made comparatively cheaply and quickly in a small mechanical workshop.

Fitted with ordinary commercial amplifiers (e.g. those from Perkin Elmer GmbH, Bodenseewerk) the detector described here can reach a sensitivity of 3×10^{-10} g/sec for packed columns as the lower limit of detection.

Properties of the Flame Ionization Detector (FID)

1. Non-temperature sensitive, practically non-sensitive towards changes in carrier gas pressure or flow rate, provided that high purity gases are used (i.e. in this context, high purity means free from hydrocarbons and other combustible substances, but contaminated with water, CO_2, and other inorganic gases and the practically non-volatile liquid phase).

2. Non-sensitive towards mechanical shock.

3. Smallest possible detector volume—less than 1 microlitre.

4. Very small time constant.

5. Extraordinarily small noise (in the order of 10^{-14} A).

The detector will not record the presence of

H_2O, H_2S

CO, CO_2, CS_2, COS, SO_2, SO_3,

CCl_4

H_2, N_2, O_2, inert gases and all other inorganic gases, except for those which are relatively easy to ionize, such as halogens, etc.

All volatile silicon compounds which do not contain carbon.

The detector will record the presence of all organic substances which form CO_2 as a final combustion product.

When properly constructed, the detector possesses a very wide dynamic range over which it can perform a linear conversion of the combustible substance provided into a stream of ions.

It will work linearly from 0–0·5 per cent substance concentration in the carrier gas, provided that the amplifier used gives linear amplification.

Because of these properties the FID may be used:

(a) for direct trace analysis with packed columns.

(b) for high temperature gas chromatography and for all chroma-thermographic methods, provided that the liquid phase used has a sufficiently low vapour pressure at the highest temperature used.

(c) as an ordinary detector for all analyses for which the presence of the substances listed above as not recorded is not of interest or indeed is a positive hindrance.

(d) as a direct measuring instrument without the use of a chromatographic column for recording the concentration of all the combustible organic substances in the concentration range 10^{-8} to 10^{-2} per cent.

Even though the FID destroys the substance by combustion it can still be used as a control instrument for preparative gas chromatography—especially for automatic operation—as well as for normal analytical gas chromatography, since it only requires 20 ml carrier gas/hr. Supposing that a preparative apparatus uses 5 l./hr carrier gas, only 0·5 per cent of this need be diverted to the detector, so that 99·5 per cent of the substance separated remains unchanged.

As the detector is non-sensitive towards all inorganic gases, it does not matter which carrier gas is used, since this will have no influence on the sensitivity of the detector (except in the case of oxygen).

The theory of the FID was dealt with by Ongkiehong[27] and fundamental studies have been published by Desty, Geach and Goldup[28] and Condon, Scholly and Averill[29]. The following data originate from work performed by the author and his colleagues Holzhäuser, Kuhl, Struppe, Hinneburg and

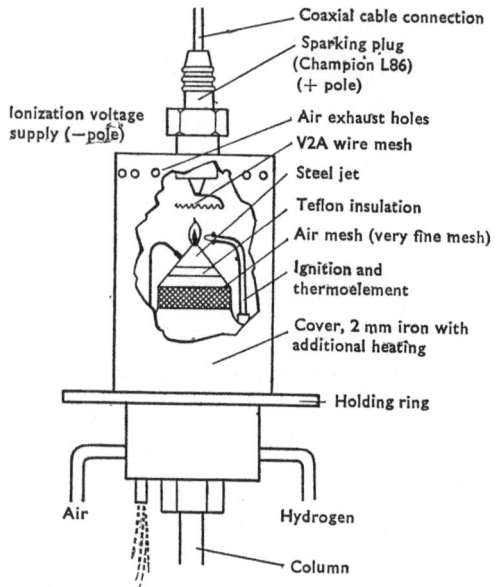

Figure 60. Method of construction of a robust flame ionization detector.

Fiedler. Since the first publication by McWilliam and Dewar (1958), an increasing interest in this detector may be noted. With the β-ray ionization detector, it has been instrumental in the development of capillary gas chromatography.

The following points should be remembered in the construction and operation of an FID (see also *Figure 60*):

(a) a sparking plug with a very high degree of insulation (e.g. Champion type L 86) should be used as the electrode holder. The distance between the electrode and the jet should be 6–10 mm;

(b) the high impedance electrode should be connected to the amplifier inlet by a highly stable coaxial cable with top quality insulation and a coaxial plug which fits the sparking plug connection;

(c) the V2A steel jet should be earthed and should be massive in order to avoid any incandescent effect; for this purpose the chamber casing can be used as insulation. It is essential that this insulation should extend also over the porcelain portion of the sparking plug. The jet should form the negative pole;

(d) the detector should be supplied with external heat to avoid any condensate separation, and the sparking plug should be kept hot. The flame of the detector should be kept hot, there should be no unecessary heat losses;

(e) only electrolytic hydrogen and high purity nitrogen (containing no more than 10^{-3} per cent vol. oxygen) or other carrier gases guaranteed free from hydrocarbons should be used (e.g. argon often contains methane, nitrogen often contains oil vapour). Even the combustion air should be purified where necessary over CuO at 750°C;

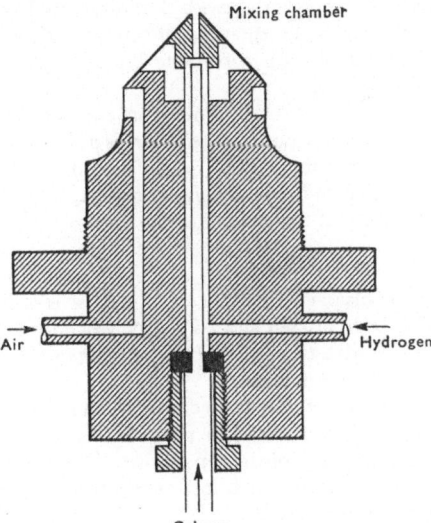

Figure 61. Assembly of the column and position of the gas supply tubes in the flame ionization detector.

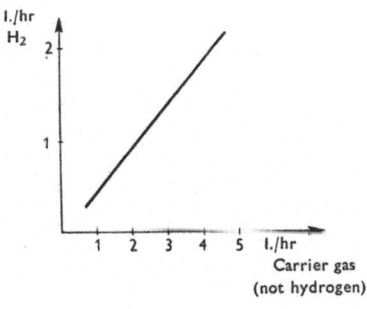

Figure 61a. Inflammability of the gas mixture.

(f) rubber tubing must not be used for the gas supply, the whole system must be fitted with annealed steel or glass tubes; the hydrogen supply tubes should have an internal diameter of 0·5 to a maximum of 1 mm. The hydrogen supply must be kept absolutely constant, either by a multistage pressure controller with a simple needle valve coupled in after it, or else by a high precision needle valve joined directly to the hydrogen cylinder. The hydrogen supply tubing must have as small a volume as possible and not be exposed to any temperature fluctuations. *Figure 61* shows the necessary gas mixing ratios in the flame.

Basic Ionization Current

The basic ionization current depends directly on the quantity of hydrogen used. The smallest flame measured on the instruments shown in the diagrams could be kept stable with 4 ml/min hydrogen. For 20 ml/min hydrogen (mixed with the quantity of nitrogen given in *Figure 61*) when connected to a column with a suitable liquid phase the basic ionization current is $1–8 \times 10^{-12}$A.

The smallest basic ionization current theoretically possible, which can arise from resistance noise alone, is about 5×10^{-15}A.

The basic ionization current should be kept as small as possible, and it is influenced quite decisively by the vapour pressure of the liquid phase and the purity of the gas used.

Noise and Drift

For this particular detector the noise is of the order of 10^{-13}A, when about 1·5 l./hr hydrogen and 2·7 l./hr nitrogen together with 30 l./hr air are being used. With a noise of 10^{-13}A it is still possible to have a lower detection limit of 10^{-10} g/sec substance.

If the noise can be reduced to the theoretically possible lower limit of 10^{-15} A then it would be possible to detect as little as 10^{-12} g/sec substance with a detector of the type shown. The drift can be kept to the same order as the noise. For such low concentrations, however, the rules and laws of ultramicro analysis must be remembered. A hypodermic syringe which has once been used to inject petroleum ether vapour will be found, even after several days use in air analysis, to be still so strongly contaminated with petroleum ether that more than 10^{-10} g of this tenacious substance are injected at each analysis. It is thus necessary to take precautions for cleanliness which are similar to those taken by a doctor in the infectious diseases section of a hospital; the only difference is that hydrocarbons are not so easy to get rid of as living biological material. The high sensitivity of the FID is naturally affected by even the smallest traces of vaporized liquid phase. This is especially troublesome when high temperature trace analysis, for which the highest possible detector sensitivity is required, is being carried out. Even the slightest traces of volatile impurities can cause a considerable disturbance under such conditions, since they produce a high basic ionization current. For this reason it is necessary in many cases to purify the liquid phase by appropriate means, e.g. by molecular distillation.

In such cases it is also found that even slight fluctuations in temperature or flow rate can cause a considerable disturbance, because they affect the quantity of vaporized liquid phase entering the detector in unit time. The detector itself is, as always, non-sensitive towards temperature or flow rate fluctuations; it is the secondary effect of the vaporized liquid phase or the impurities therein which causes the disturbance.

116

Further details on these and other problems, particularly on points related to the ionization current amplifier, will be found in Volume II.

Ionization Voltage

The FID has a saturation voltage, above which an increase does not give any further current output. At the highest concentrations the saturation current may only be reached at around 200 volts d.c. At ionization voltages above 300 V a further current output is possible (secondary effect), but only by a deterioration of the signal-noise ratio, which, of course, should be as high as possible. Normally an ionization voltage of 100–200 volts (d.c.) is used. Details on the use of a.c. with the FID will be found in Volume II.

Response Factor

Equal weights of hydrocarbons produce equally large deflections. Equal mole quantities of hydrocarbons produce deflections that are directly proportional to the carbon content of the molecule.

Each atom of oxygen, nitrogen, sulphur, halogen, etc., in the molecule decreases the deflection in a manner which corresponds not only to the weight percentage decrease of the carbon content of the molecule, but generally to an even greater extent according to the structure of the hetero-atom bonding.

If a FID analysis is to be evaluated quantitatively—which is well within the bounds of possibility, thanks to the wide dynamic range of $1 : 10^6$ and the good linearity—the following formula may be used to correct the peak area values obtained to weight percentage ratios:

$$f = \frac{\text{Mole weight}}{\text{Number of carbon atoms} \times 12}$$

This correction factor is only an approximation, which is shown by the following example:

The three isomeric xylenes, which according to the above formula should possess completely equal factors, show up to 5 per cent difference in the peak areas in 33·3/33·3/33·3 per cent wt. mixtures.

More exact correction factors will be found in Volume III (Tables).

On account of the above-mentioned properties, the FID can be used in combination with the gas density meter or the thermal conductivity cell for identification purposes.

Remember that the suppression of the sensitivity of the FID necessary for such comparisons can cause the amplifier belonging to it to operate non-linearly!

As the FID, together with the β-ray ionization detector, is chiefly used as a detector for capillary gas chromatography, further practical data, together with the necessary theoretical data for both these detectors, will be found in Volume II. Data on ionization current amplifiers will also be found there.

Titration Detectors and Volumetric Detectors
(Janak Principle)

Both of these types of detectors are, in comparison with differential detectors, not very sensitive, and in addition to this of only limited application. For special tasks, such as the analysis of acids or bases and the analysis of gases which are not soluble in carbonic acid, they have certain advantages. For example, the titration detector gives the equivalent of bases or acids as a sum, and the correctness of its reading depends practically only on the correct choice of the titration alkali or acid.

James and Martin[19] give such detailed data on their titration integral detector that it would certainly be possible to construct one if required. A good integral detector which records the volume/time curve and works by the Janak principle was described by Leibnitz, Könnecke and Hrapia[20].

Nowadays trace analyses in particular are carried out in such a way that the organic components are burned to CO_2 and H_2O over CuO or CoO. In this way the detectors which obtain a value for CO_2 can be regarded as detectors of the first type. A very high sensitivity may thus be obtained if the carbonic acid is not titrated, but instead its effect on an indicator solution is measured optically. (See also under the section on apparatus.) CO_2 detectors have been described by Liberti[21] and Boer [22]; they are similar in principle to the titration detector of James and Martin[19].

Testing of Detectors

Detectors should be tested for:

1. their linear signal over the concentration range 0 to 10 per cent vapour in the carrier gas; for their sensitivity, measured as limit of detection;
2. their specific response towards different types of chemical compound and increasing molecular weight within a homologous series; this is useful in comparative analysis, where the content of an always identical reference substance, e.g. benzene or pentane or methyl stearate, in the mixture is exactly known;
3. their signal delay or their half life period and dead volume;
4. their susceptibility towards:
   (a) variations in flow rate;
   (b) variations in temperature;
   (c) variations in pressure;
   (d) variations in the operating conditions of the detector (e.g. the voltage).

If the influence of 1 to 4 for an amount of added substance which is always identical is studied on the absolute and relative quantitative result—for this the sum of the peak areas of the components is measured as the absolute result, and the relative peak areas of, for instance, pentane and pentanol or benzene and ethanol, are also measured—then we can obtain the maximum limits of variation of the operating conditions which must be maintained if a given degree of precision is required in the quantitative analytical result.

We shall conclude by stating once again: the type and operating conditions of the detector have no influence on the qualitative result, but always have an influence on the quantitative analytical result.

Comparison of Sensitivity

In order to be able to obtain a quantitative comparison of the different types of detector it is expedient to specify a characteristic value which will be equally suited to all the different principles of measurement. The function of the detector should be remembered: it has to convert an unknown substance in the carrier gas stream into a value which can be recorded e.g. a concentration is converted into mV. The author proposes the following characteristic for detectors.

The so-called corrected sensitivity E is defined as the ratio of the voltage (relative to the noise and drift) to the concentration of substance in the carrier gas passing through the detector:

$$E = \frac{U_{max.}/R + D/h}{mg_{max.}/ml}$$

$U_{max.}$ = minimum voltage at the recorder in mV; R = noise in mV; D/h = drift in mV/hr; $mg_{max.}$ = quantity of substance (in mg) flowing through the detector at the instant of the maximum electrical deflection; ml = the quantity of carrier gas under normal conditions flowing through the detector in the same space of time.

This definition presupposes that all the different types of detector are to be tested with the same substance, as only a few of the principles of measurement give values which are independent of the substance and the carrier gas. It is therefore expedient to use, for example, methane, benzene, or methyl stearate as the test substance and to specify the carrier gas. An advantage of the definition is that the maximum output of the detector must be related to the quality of the electrical signal. For this it does not matter whether electronic amplification is used or not, i.e. the values obtained represent truly comparable values.

Calculation of E

The maximum deflection $U_{max.}$ is replaced by h(mV) (peak height). The maximum concentration $C_{max.}$ is obtained from the following relationship.

Let the total amount of substance injected be e. It produces a chromatographic signal of area h (mV) . $b_{\frac{1}{2}}$ (min). The maximum concentration produces the deflection h and should prevail on the average over the very short time period Δb.

The quantity of substance flowing during this time is $e_{max.}$. It corresponds to the area $h . \Delta b$.

The amount of carrier gas flowing during the time b is x ml.

$$\frac{x}{\Delta b} = \frac{F_c . 1,000}{60} \qquad\qquad x = \frac{\Delta b . F_c . 100}{6}$$

F_c = L/h carrier gas at the detector temperature c.

Thus the maximum amount is

$$\frac{h . \Delta b}{e'_{max.}} = \frac{h . b_{\frac{1}{2}}}{e} \qquad\qquad e_{max.} = \frac{h . \Delta b . e}{h . b_{\frac{1}{2}}} = \frac{\Delta b . e}{b_{\frac{1}{2}}}$$

The maximum concentration C_{max}. is

$$C_{\mathrm{max.}} = \frac{\frac{\Delta b \cdot e}{b_{\frac{1}{2}}}}{\frac{\Delta b \cdot F_e \cdot 100}{6}} = \frac{e \cdot 6}{F_e \cdot 100 \cdot b_{\frac{1}{2}}}$$

From this we obtain

$$E = \frac{\frac{h}{R + D}}{\frac{e \cdot 6}{F_e \cdot 100 \cdot b_{\frac{1}{2}}}} = \frac{h \cdot b_{\frac{1}{2}} \cdot F_e \cdot 16 \cdot 66}{e(R + D)}$$

In the literature another method is given for reporting the comparative sensitivity of different types of detectors: Dimbat, Porter and Stross[11] define the sensitivity S as the area in cm^2 which is produced as a peak by 1 mg substance using a 'unit recorder'. This method of classification gives the following general picture:

Detector	S in $mV \cdot ml./mg$
Micro flame ionization detector	5×10^9
Thermal conductivity cell with Pt/Ni ribbon wire $4 \times 40\mu$ using hydrogen	2×10^3
Beckman (USA) thermal conductivity cell with hydrogen	1×10^3
Hartmann und Braun (Frankfurt am Main) thermal conductivity cell with glass-covered measuring element, using hydrogen	800
Ionization detector with ^{90}Sr and ultrapure helium	5×10^{11}
Ionization detector with ^{90}Sr and pure helium	600
Scott micro flame detector for hydrocarbons	200
Grant flame detector for hydrocarbons	12
Simple thermal conductivity cell with 30μ Pt wire, 10 ohm, hydrogen	20
Lovelock ionization detector with ^{90}Sr and argon as carrier gas, for hydrocarbons	3×10^{10}

Figure 62 shows how the value for E is obtained. Especially in the construction of a detector and also in the selection of a commercial instrument it is valuable to check the sensitivity of the instrument by the determination of E.

If the two methods for determining the detector sensitivity given above are compared it may be seen that the method of Dimbat, Porter and Stross possesses one disadvantage, namely that the values of S do not give the true quality of the detector, but are dependent upon the operating conditions prevailing at that time (*Figure 63*).

Thus in contrast to E, S is not a characteristic value (see also Holzhäuser[12]).

If the equation for E is compared with the equation for S it can be seen that they differ only by a factor of $\dfrac{1}{R + D}$.

This is, however, quite considerable, in so far that only one value for E is possible as opposed to any number of values for S, and further that the limiting value for S is given relative to the electrical, thermal and gas kinetic qualities of the detector. The data are, however, still dependent on the type of carrier gas and the substance.

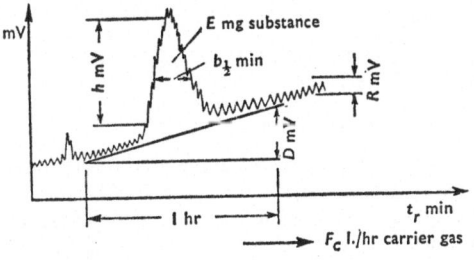

Figure 62. Determination of the detector sensitivity E.

A further characteristic for sensitivity is the so-called lower limit of detection. It is defined as the amount of substance in grams per ml carrier gas or grams per second capable of producing a detector or recorder deflection of 200 per cent of the noise, in other words a deflection just capable of being recognized, and is abbreviated to LS (lower sensitivity limit). We may define the noise as the temporary and irregular fluctuation of the base line (maximum value) in mV. Now suppose that we produce a stream of carrier gas saturated with the substance at constant temperature. Under such circumstances we can calculate the exact number of grams of substance which are

Figure 63. Determination of the detector sensitivity S according to Dimbat, Porter and Stross.

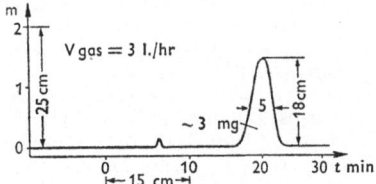

being carried through the detector per second, and we may call this amount a g/ml gas. In the same way we can find out exactly how many ml carrier gas per second are flowing through the detector, and this amount is F ml/sec. Under such circumstances the deflection in mV is I

$$LS = \frac{\rho . F . 2 . R}{I} \text{ g/sec.}$$

R = noise in mV
ρ = concentration of substance in gas in g/ml
F = gas flow through detector in ml/sec
I = ionization voltage at detector in mV produced by $\rho . F$ g/sec substance.

Example:

Noise: 0·1 mV for full sensitivity (no shunt)
Flow of substance: 1×10^{-6} g/ml through the detector at 0·5 ml carrier gas/sec;
Suppose that this substance flow produces a detector reading of 6·6 mV at a sensitivity of 1/125 full sensitivity.

Calculation based on definition:

$$LS = \frac{0·5 \cdot 2 \cdot 1 \cdot 10^{-6} \cdot 0·1}{125 \cdot 6·6}$$

$$= 1·22 \times 10^{-10} \text{ g/sec.}$$

Such a figure is of practical value. Suppose, for example, that it is required to know if a detector is suitable for trace analysis; the value for *LS* will show how many ml substance must be injected in order to produce a measurable deflection for the trace substance which can clearly be distinguished from the noise.

References

1. DEAL, C. H., OTVOS, J. W., SMITH, V. N. and ZUCCO, P. S., *Analyt. Chem.*, 1956, **28**, 1958.
2. BOER, H., *Vapour Phase Chromatography*, ed. D. H. Desty, Butterworths, London, 1957, p. 169.
3. LOVELOCK, J. E., *J. Chromatog.*, 1958, **1**, 35.
4. MCWILLIAM, I. G. and DEWAR, R. A., *Gas Chromatography*, ed. D. H. Desty, Butterworths, London, 1958, p. 142; *Analyt. Chem.*, 1957, **29**, 925.
5. GRANT, D. W., *Gas Chromatography*, ed. D. H. Desty, Butterworths, London, 1958, p. 153.
6. RYCE, S. A. and BRYCE, W. A., *Nature, Lond.*, 1957, **179**, 541; *Ref. Chem-Ing-Tech*, 1957, **29**, 486.
7. SCOTT, R. P. W., *Vapour Phase Chromatography*, ed. D. H. Desty, Butterworths, London, 1957, p. 131.
8. National Instr. Lab. Inc., Riverdale, Md, *Chem. Engng*, 1957, **64**, 3, 192.
9. GRIFFITHS, J., JAMES, D., and PHILLIPS, C., *Analyst*, 1952, **77**, 897.
10. MARTIN, A. J. P. and JAMES, A. T., *Biochem. J.*, 1956, **63**, 138.
11. DIMBAT, M., PORTER, P. E. and STROSS, F. H., *Vapour Phase Chromatography*, ed. D. H. Desty, Butterworths, London, 1957, under Boer, H., p. 169.
12. HOLZHÄUSER, H., diploma dissertation, Karl-Marx University Leipzig, (1958), Faculty of Mathematics and Natural Sciences.
13. BOTHE, H. K., *Gas-Chromatographie 1958*, ed. H. P. Angele, Akademie-Verlag Berlin (1959).
14. WELLMANN, W. L. and LOVELOCK, J. E., *J. Instn Heat. Vent. Engrs*, 1955, **22**, 421.
15. DIMBAT, M., PORTER, P. E. and STROSS, F. H., *Analyt. Chem.*, 1956, **28**, 290.
16. KEULEMANS, A. I. M., *Gas Chromatography*, Reinhold Publ. Corp., New York, 1957, p. 79.

References

17. KAISER, R., *Gas-Chromatographie 1958*, ed. H. P. Angele, Akademie-Verlag, Berlin, 1959, 327.
18. LEIBNITZ, E., KAISER, R. and HOLZHÄUSER, H., *Gas-Chromatographie 1958*, ed. H. P. Angele, Akademie-Verlag, Berlin, 1959, 271.
19. JAMES, A. T. and MARTIN, A. J. P., *Analyst*, 1952, **77**, 915.
20. HRAPIA, H., *Gas-Chromatographie 1958*, ed. H. P. Angele, Akademie-Verlag, Berlin 1959, p. 144.
21. LIBERTI, A., *Analyt. chim. acta*, 1957, **17**, 247.
22. BOER, H., 'The Use of Ozonolysis. . . .', *Proc 4th World Petrol. Congr.*, Sect. VA, paper 1, Rome, 1955.
23. LIBERTI, A., COSTA, G. and PAULUZZI, E., *Chim. e Industr.*, 1955, **38**, 674.
24. PRIESTLEY, W., *Vapour Phase Chromatography*, ed. D. H. Desty, Butterworths, London, 1957, p. 165.
25. THOMPSON, A. E., *J. Chromatog.*, 1959, **2**, 148.
26. ANDREATCH, A. J. and FEINLAND, R., *Analyt. Chem.*, 1960, **32**, 1021.
27. ONGKIEHONG, L., *Gas Chromatography 1960*, ed. R. P. W. Scott, Butterworths, London, p. 7.
28. DESTY, D H , GOLDUP, A. and GEACH, C. J., *Gas Chromatography 1960*, ed. R. P. W. Scott, Butterworths, London, p. 46.
29. CONDON, R. D., SCHOLLY, P. R. and AVERILL, W., *Gas Chromatography 1960*, ed. R. P. W. Scott, Butterworths, London, p. 30.
30. SCHMAUCH, L. J., and DINERSTEIN, R. A., *Analyt. Chem.*, 1960, **32**, 343.

Further Literature on Detectors

BENNETT, C. E., NOGARE, S. D., SAFRANSKI, L. W. and LEWIS, C. D., *Analyt. Chem.*, 1958, **30**, 898.
COWAN, C. B. and STIRLING, P. H., The selection and operation of thermistors for katharometers, *Gas Chromatography*, ed. Coates, Noebels, Fagerson, Academic Press Inc, New York, 1958, p. 165.
DAVIS, A. D. and HOWARD, G. A., Use of thermistor detectors in gas chromatography, *Chem. & Ind.*, Brit. Inds. Fair Rev., 1956, R25.
DAVIS, A. D. and HOWARD, G. A., Thermistor detectors in gas chromatography, *J. Appl. Chem.*, 1958, **8**, 183.
DESTY, D. H., Vapour detectors for gas chromatography, *Nature, Lond.*, 1957, **180**, 22.
EBEID, E. M. and MINKOFF, G. J., Sensitivity of Detectors, Res. Corresp.: Suppl. to *Research*, 1956, **9**, 24.
FELTON, H. R. and BUEHLER, A. A., High temperature thermal conductivity cell (for temperatures up to 500°C with model airplane 'glow plug' as the measuring element), *Analyt. Chem.*, 1958, **30**, 1163.
GLUECKAUF, E., BARKER, K. H. and KITT, G. P., Detectors for radioactive substances in GLC, *Disc. Faraday Soc.*, No. 7, 1949, 199.
GREEN, G. E., A hydrogen-conversion detector for gas chromatography, *Nature, Lond.*, 1957, **180**, 295.
GROSSKOPF, K., Testing tubes as detectors for gas chromatography, *Erdöl u Kohle*, 1958, **11**, 304.
HARLEY, J., *et al.* Flame ionization detector for gas chromatography, *Nature, Lond.*, 1958, **181**, 177.
HARLEY, J., and PRETORIUS, V., A new detector for vapour phase chromatography, *Nature, Lond.*, 1956, **178**, 1244.

HENDERSON, J. I. and KNOX, J. H., The micro flame detector in gas-liquid partition chromatography: correlation of response with heats of combustion, *J. Chem. Soc.*, 1956, 2299.

HESELTINE, H. K., PEARSON, J. D. and WAINMAN, H., A simple thermal conductivity meter for gas analysis with special reference to fumigation problems, *Chem. & Ind.*, 1958, 1287.

JOHNSTONE, R. A. W. and DOUGLAS, A. G., A detector for gas analysis (UV absorption), *Chem. & Ind.*, 1959, 154.

LIPSKY, S. R., LANDOWNE, R. A. and LOVELOCK, J. E., Separation of lipides by gas-liquid chromatography (argon ionization detector), *Analyt. Chem.*, 1959, **31**, 852.

MADAN, M. P., Simple bridge method for the measurement of thermal conductivity of gases and gas mixtures, *J. Franklin Inst.*, 1957, **263**, 207.

MURRAY, K. E., New design of the Martin and James gas density meter, *Aust. J. Appl. Sci.*, 1959, **10**, 156.

NEDOROST, M., Polarographs as detectors, *Chem. Listy*, 1956, **50**, 317.

TURNER, D. W., A robust but sensitive detector, for gas-liquid chromatography, *Nature, Lond.*, 1958, **181**, 1265.

WALKER, R. E. and WESTENBERG, A. A., *Rev. sci. Instrum.*, 1957, **28**, 789.

YOUNG, I. G., Survey of detectors, ISA Symposium on Gas Chromatography, Michigan State University, 1959.

GAZIEV, G. A. and YANOVSKII, M. I., A radiometric detector for the analysis of radioactive gases in GSC, *Izdatel'stvo Akad. Nauk SSSR*, 1957, 405.

Further references will be found in the volumes of *Gas Chromatography Abstracts* ed. C. E. H. Knapman, Butterworths (1958, 1959, 1960 and following) under Subject Index No. 3.4.

2.5. RECORDERS

A LARGE number of chromatographic fractions leave the column in rapid succession: the rise and fall of a peak may be over in a matter of seconds. The amounts of substance involved are very slight, and the voltages produced by the detector are thus often very small, being generally of the order of a few millivolts.

If in a gas chromatographic analysis the precision is required to be ± 0·2 per cent (and this can be achieved) and the readability of a line on the chart must be 0·5 mm, the width of the paper strip must be 200 mm.

For such a precision the reproducibility of the deflection, the linearity of the scale, and the constancy of the chart speed must also be kept in the region of ± 0·2 per cent.

Such requirements may be met only by an electronic compensograph. For normal analyses electronic continuous line recorders are used. These have a recording width of 200–250 mm and a scale value of 0–1 mV (also 0–10 mV) and must have a scale passage time of less than 1 sec—in high speed analysis even less than 0·5 sec.

The input resistance values conform to the detector or amplifier being used. They are in the region of a few thousand ohms or less than 100 ohms.

In order to exploit to the full the accuracy of recording and reading of the recorder and in order to be able to obtain quantitative results over as wide a dynamic range as possible, one of the following methods should be used.

1. Two recorders are used, connected in parallel with the detector. Of these, one works at maximum sensitivity, for which the noise can be *ca.* ± 0·5 per cent of the scale value, and the other works at a sensitivity which has been diminished so much that even the largest deflection still remains within the region of measurement.

2. A recorder with automatic measurement range change-over is used. When the recording pen reaches 95 per cent of the full deflection it operates a contact which switches it by a constant factor to a lower sensitivity. If on the return journey the pen reaches 5 per cent of the full deflection the reverse process happens, i.e. it is switched to the next highest sensitivity.

3. A simple recorder with a small scale width (e.g. 100 mm) and a logarithmic scale is used. Such a recorder registers lower concentrations as peaks which can still be clearly seen, but is also able to record the maximum possible value. If quantitative analytical results are wanted, then an integrator must be connected directly to the detector (or to the output of the amplifier) so that the linear integrals can be obtained and printed out or recorded.

It is in practice very useful if a simple test bench can be constructed, on which an electronic compensograph can be tested and adjusted to give optimum results. Rate of carriage transit and damping may well vary over a long working period. Further, the earthing, insulation and adjusting device must be functioning in an optimum fashion. In this way it is easier

Connection with compensograph

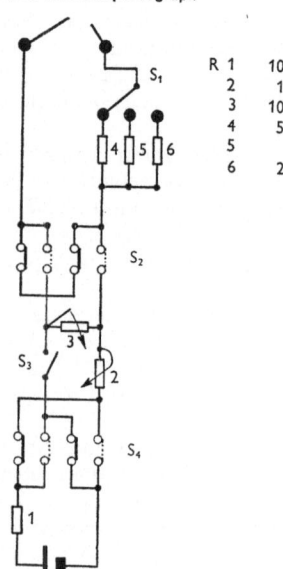

R 1	100 Ω
2	10 K Ω
3	100 Ω
4	50 Ω
5	1 K Ω
6	20 K Ω

Figure 64. Test arrangement for the adjustment of sensitivity and control of suitability of the recorder.

to find and correct faults. The arrangement shown in *Figure 64* is recommended.

2.6. TEMPERATURE

The following section deals with the influence of the column temperature on the analytical results, and describes the theoretical and constructional basis for temperature control in gas chromatography.

IN THE practice of gas chromatography the operating temperature plays a decisive part. While it is normally important in GLC to keep the temperature as constant as possible during the analysis and to adjust it to an identical value for each analysis, in GSC it is preferable to work with progressive or linearly increasing temperatures or temperature cycles.

Even in GLC, many problems which can only be solved with difficulty by means of isothermal GLC can be dealt with in a simpler and more advantageous manner by varying the temperature in a progressive or linear fashion during the analysis. Of course this increases the amount of apparatus required considerably, and there is a whole series of additional factors that must be taken into account when working with the non-isothermal procedure. This will be referred to again.

It has been found experimentally that the specific retention volume V_g decreases exponentially with rising temperature, or, expressed more exactly:

$$\log V_g \sim \frac{1}{T} \qquad \ldots.(1)$$

From thermodynamical considerations the following equation may be obtained (Ambrose et al.[1])

$$\log V_g = \frac{\Delta H_s}{2 \cdot 3 \cdot R \cdot T_s} + c \qquad \ldots.(2)$$

V_g = specific retention volume (see also Section 1); ΔH_s = partial molar heat of evaporation of the dissolved substance from the solution; $R = 1{,}987$ cal/degree mole, gas constant; T_s = temperature of the column in °K; c = constant.

If equations 1 and 2 are compared, it can be seen that ΔH_s should remain constant with changing temperature provided that $\log V_g \cdot T$ remains really constant. This is, however, not quite the case, and equation 1 is of only limited validity. However, within a homologous series the variations of the values for ΔH_s with temperature for the different homologues are very similar, and thus in this case we may say that the relative corrected retention volume is truly inversely proportional to the temperature T;

$$\log \frac{V_g}{V_{gB}} = \log r_B = c \cdot \frac{1}{T_s} \qquad \ldots.(3)$$

see also *Figure 65*.

V_{gB} = specific retention volume of a reference substance
r_B = retention relative to a reference substance
c = constant.

From equation 3 it can be seen that the chromatogram of a homologous series made with linear increase in temperature will show a time sequence for the individual homologues. The peaks thus no longer appear with a logarithmic separation. In this way the time of analysis can be shortened 'logarithmically'.

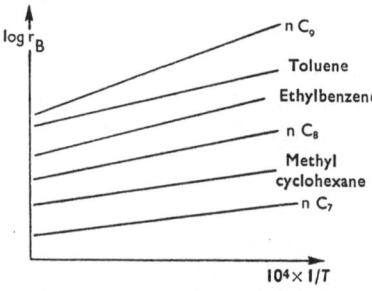

Figure 65. Dependence of retention r_B on the temperature T.

However, it is only very rarely that such simple mixtures as pure homologous series are met with in practice. Moreover the number of isomers or compounds of similar structure is known to increase rapidly with increasing molecular weight, so that isothermal GLC is to be preferred. In spite of this the use of step increases in temperature in particular is very often useful. Basically there are three methods of exploiting the time-shortening temperature factor:

1. The column is run with linear increase in temperature.
2. The column is run stepwise and isothermally.
3. Two or more columns are used for an analysis, each at a different temperature. The sample is introduced into column 1, which is at the highest temperature. The rapidly travelling components which are poorly separated if at all are washed into the second, colder column, and so on. Meanwhile the more slowly travelling components are still being separated in column 1. The columns thus operate in parallel and are only connected in series for a short time.

Method 3 is the best solution to the problem. It combines the advantages of isothermal gas chromatography with the advantages of the temperature programmed mode of operation. The apparatus required is, however, just as complicated as that for multi-column instruments. The amount of apparatus required is fairly large.

Figures 66 and *67* show the comparison between isothermal gas chromatography and the non-isothermal methods 1 and 2.

Greene *et al.*[2] and Evans and Willard[3] have, among others, investigated the influence of temperature on the separability of difficult mixtures, and even in 1952 Griffiths *et al.*[4] were indicating the advantages of chromathermography. In 1954 Berridge and Watts[5] published a paper on the advantages of increasing column temperature in the separation of homologous series.

In Germany Kögler[6] in particular has concerned himself with the problems of chromathermography, as the non-isothermal procedure may be briefly described.

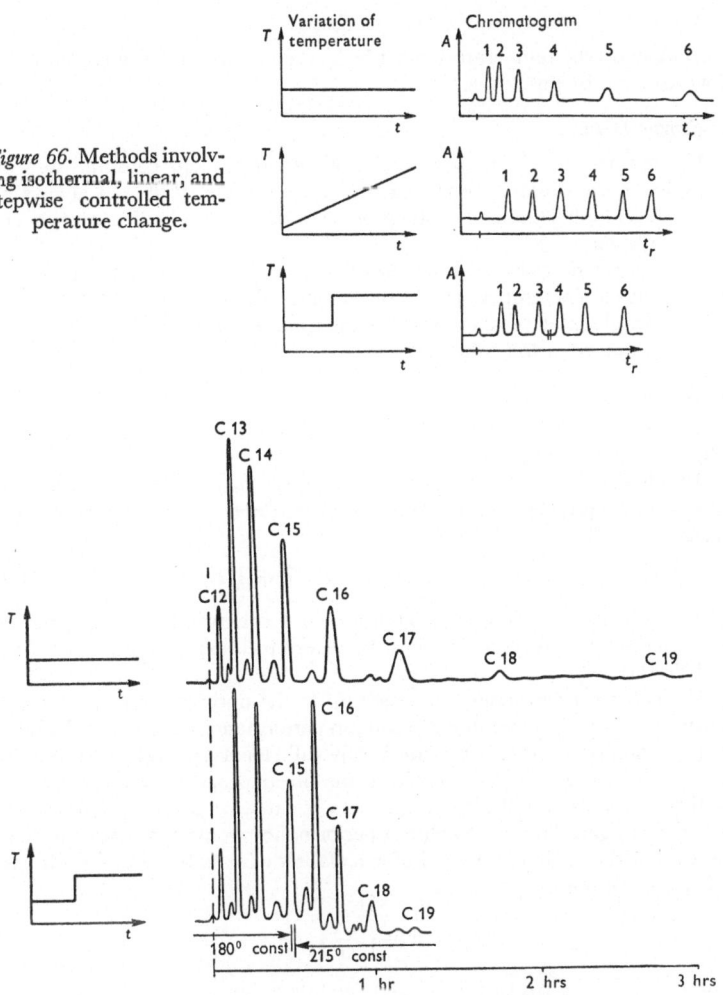

Figure 66. Methods involving isothermal, linear, and stepwise controlled temperature change.

Figure 67. Saving of time in chromathermography.

Kamer *et al.*[7] and Ashbury *et al.*[8] dealt with the technique of progressive temperature increase, and the difficulties arising in the application of

129

chromathermography were described by, among others, Lichtenfels[9] and Patton[10]. Detailed constructional data are given by Harrison *et al.*[11].

The different effects of the operating temperature on the individual parts of the gas chromatograph are as follows:

1. *Gas Control, Measurement and Control Systems*

In all cases the temperature must be kept constant, so that a constant gas flow rate can be guaranteed.

2. *Sample Injector*

The temperature here must on the one hand be kept constant, and on the other must conform to the nature of the sample. It should always be about 30°C higher than the boiling point of the highest-boiling component of the sample.

It is known that the peak height of a substance is within certain limits dependent on the temperature of the sample injector (Section 2.3).

The sample injector may be at a temperature similar to, or higher than, the temperature of the column.

3. *Column*

(a) The primary influence of the temperature is on the retention values. The influence is logarithmic.

(b) However, the temperature also affects the column performance. The height equivalent to a theoretical plate depends on the temperature according to the equation

$$HETP = a + b \cdot T + c/T \qquad \ldots(4)$$

where a, b, and c are constants. Thus for every column there is an optimum operating temperature, which must be determined experimentally (de Wet and Pretorius[12]).

The column performance increases up to the optimum value, then falls away. It should be noted that the column performance cannot be calculated by the ordinary equation, since this is only valid for isothermal gas chromatography. As the width of the peaks in the linear programmed temperature method remains practically constant (a great advantage, which will be referred to again later) the column performance would increase with temperature if the ordinary method of calculation were to be used. In this case it is better to use the equation

$$n_{\frac{1}{2}} = 5 \cdot 54 \left(\frac{t_{drT}}{b_{\frac{1}{2}}}\right)^2 \cdot \frac{1}{L}$$

where t_{drT} = temperature corrected retention time of the substance with peak width $b_{\frac{1}{2}}$, which leaves the column at a retention temperature of T. Temperature correction is here taken to mean that the retention time given is that which the substance would have had if the column had been run isothermally at the temperature T (L = length of column in m). The rise in column performance is thus only a formal one, and the peaks only remain approximately equal in width as long as the temperature is increasing. It is, however, possible to reduce the peak width of a substance from, e.g. 4 min when run isothermally at 168°C to 1·5 min at 236°C, although for both analyses the same time of 28 min is required. This, however, increases the peak height for the substance under consideration.

(c) At the same time the chromatogram from a temperature programmed analysis is capable of considerably better evaluation than that from an isothermal analysis carried out under comparable conditions.

(d) The temperature can sometimes exert a considerable effect on the selectivity of the column. It is known that the partition coefficient K is also temperature dependent according to the equation

$$\log \frac{K}{T . \rho L} = \frac{\Delta H_s}{2 \cdot 3 . R . T} + C \qquad \text{....(5)}$$

(see also Ambrose[1]).

where ρL = density of the liquid phase and R = gas constant (1,987 cal./ g.mol.). Thus even within the same group of chemical compounds the separability of a given mixture can vary strongly with temperature. This

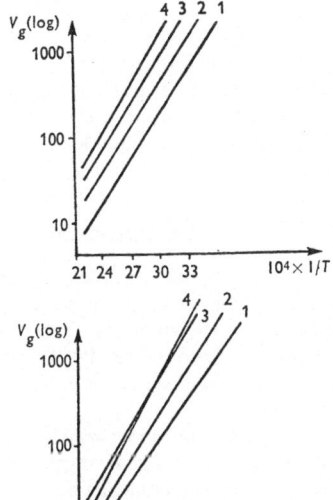

Figure 68. Variation in the temperature dependence of the retention time of different substances on varying liquid phases.

may go to the extent that at lower temperatures substance A appears before substance B, at higher temperatures A appears after B, and there may be an intermediate temperature so unfavourable that separation is completely out of the question, no matter how long the column.

It may happen that if a different stationary phase is used these phenomena are not found, in other words the operating temperature plays no part in the separation of A and B. *Figure 68* gives an example.

For this reason retention values should always be measured at at least two temperatures, and the appropriate temperature should be quoted with the V_g value, or it may be expedient to use the form

$$\log V_g = A + \frac{B}{t + C} \qquad \qquad ....(6)$$

A, B, C = constants

t = temperature in °C

and determine the values of the constants, A, B and C. In this way the relationships between the V_g values and the operating temperature are given to a sufficient degree of precision, and the V_g values can generally be evaluated. With such values the user can, among many other things, perform the following: (with progressive temperature control) the analysis of a mixture of, say, n-alkanes runs normally; it is known up to which homologue the mixture is to be investigated. By means of the equation

$$\log V_g = A \cdot n$$

(A = constant; n = carbon number) it is possible during the course of the reaction to calculate graphically or arithmetically by what time the last homologue should appear. If the time is too long it is possible by means of equation 6 to calculate the value to which the temperature must be adjusted so that the last homologue leaves the column within a given required time.

The Optimum Operating Temperature

It is not possible to give a simple relationship enabling the optimum operating temperature to be calculated from easily obtainable data. All that can be given are some rules and the possibility of calculating the suitable temperatures from the results of a practical test at a previously selected temperature from the equations in Section 1—retention temperature, equations 37 and 38.

An example may show why it is impossible to give a simple relationship.

A mixture of nitrogen, oxygen, hydrogen and methane is readily separable on Linde molecular sieve 5A at 20°. If, on the other hand, it is necessary to separate argon from oxygen (a problem which always becomes acute when a definite value for the oxygen content is to be obtained) it is necessary, with the same column packing, to use a separating temperature of -70°C.

The rules are as follows:

1. The optimum operating temperature depends on the separation and the packing chosen. The operating temperature should be high enough for the liquid phase used to have as low a viscosity as possible, but at the same time no vapour pressure worth bothering about (i.e. a maximum of 1 mg

vapour per litre carrier gas when non-sensitive detectors are used, and 10γ vapour per litre carrier gas when ionization detectors are used). The selectivity required must be fully effective at the temperature chosen.

2. The optimum separation temperature is anything up to a maximum of 100°C below the average boiling point of the sample for narrow boiling range mixtures where normal amounts are being used, and up to a maximum of 200°C below the boiling point of the highest boiling component when only gamma quantities are being injected and ionization detectors are being used. In this case the temperature of the sample injector should be up to 100°C higher than the column temperature.

3. The separation temperature may be in the region of the boiling point of the components. In this case the substances have, on a normal 1 m long GLC column, retention times of 1–2 minutes. In GSC the operating temperatures are about 200°C higher than for the same separation in GLC.

Constant Temperature

The previous section dealt with the particular influence of temperature.

As the retention values, which are of importance in qualitative analysis, vary logarithmically with the temperature, in the isothermal method the temperature constancy must be better than $\pm$ 0·1°C if truly comparable results are to be obtained. In the non-isothermal method the initial temperature value and the temperature-time gradient must also be adjustable to the same reproducibility of $\pm$ 0·1°C.

It must be possible to use the detector and the sample injector at quite definite, and reproducible temperatures.

In order to make use of the whole range of possible applications it is thus necessary to have temperatures of 0° to 400°C. A flexible laboratory instrument should therefore have an operating range of 0° to 300°C. A high temperature instrument should be usable up to 500°C.

Temperature Control

Of the various temperature control methods for gas chromatographs, only two have found general acceptance:

1. *Temperature Control with a Gas Thermostat*

Air is used as the temperature control medium. Because of its very low specific heat it needs to be circulated with extraordinary force.

2. *Temperature Control with a Metal Thermostat*

Relatively large masses of, for example, aluminium are kept heated to a constant temperature. The column and detector are located either within the externally heated metal mass, or if they are attached to the outside have the greatest possible surface contact with the thermostat metal. In both cases the thermostat must be protected from external temperature disturbances by a really good heat insulation. It is important that if possible no metal part, unless its cross-sectional area is very small, should lead from the thermostat region to the exterior. Doors, etc., are therefore made of heat-insulated material, metal surfaces which lead to the exterior are interrupted, for instance, by asbestos strips. An example of an instrument whose temperature is controlled in this manner is shown in *Figure 69*.

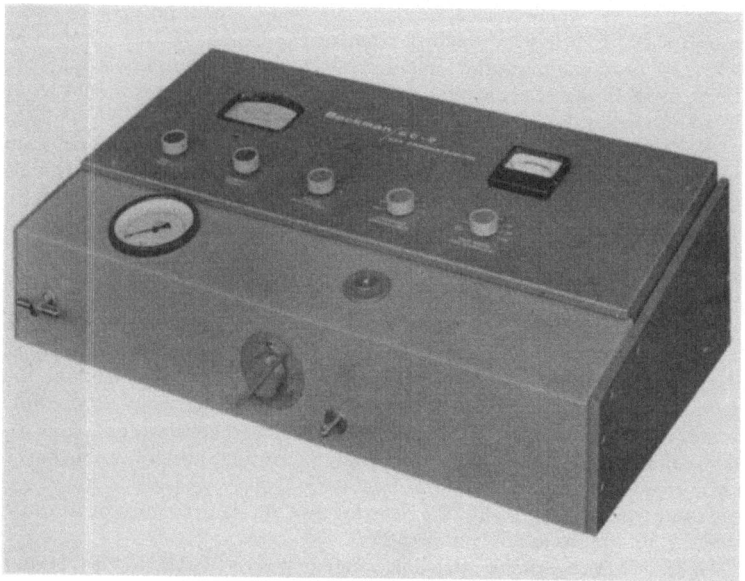

Figure 69. Gas chromatograph produced by Beckmann Instruments GmbH, Munich.

For lower temperatures (below 50°C) neither air nor metal thermostats are suitable unless they include not only a heating but also a cooling system (e.g. water cooling). In actual fact the only way to carry out precision measurements in this temperature region without a lot of expense is by means of a liquid thermostat.

Metal thermostats can only be used for the isothermal method; on the other hand air thermostats are a prerequisite for the temperature programmed method.

The Air Thermostat

The very first commercial gas chromatographs used air thermostats for temperature control.

Advantages. The part of the instrument concerned can be raised to a high operating temperature within a very short time. At an average heating rate of 15°/min an operating temperature of 200°C can be reached within 10 min, and one of 300°C within 20 min.

Because of this high heating rate, air thermostats are especially suitable for chromathermography, as practically any temperature-time programme required can be carried out.

Besides the disadvantages enumerated below, air has the advantage as a temperature control medium that it is available at any operating temperature. The low viscosity enables high circulation rates to be obtained.

Disadvantages. Air has a low specific heat. Heat losses at points where of necessity large metal surfaces must be exposed can therefore only be avoided by a very powerful circulation of the thermostat air. For this, air flow rates of up to 20 m/sec are necessary.

To control the temperature of a space 1 m long of volume 7 l. with a maximum of 500 watts steady heat loss, a circulation of up to 3 m³/min is required. A preparative gas chromatograph (see *Figure 70*) with a tempera-

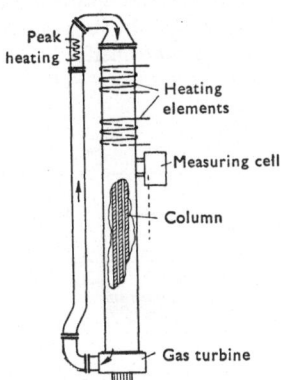

Peak heating

Heating elements

Measuring cell

Column

Gas turbine

Figure 70. Preparative gas chromatograph with air thermostat.

ture control volume of 15 l. and a linear expansion of about 1·8 m must be provided with an air circulation of 20 m³/min. In this case the temperature difference between the highest and lowest points of the thermostat is less than 2°C at 200°C. The steady heat loss is still about 500 watts.

It is known that with increasing linear expansion the circulation of the temperature control medium must be increased greatly if considerable heat losses are to be avoided. In the final analysis this requires blowers that work well and give a high output, and also are capable of withstanding high temperatures.

The author and his colleagues solved the problem by setting up a motor running at about 3,000 to 8,000 r.p.m. outside the thermostat, the motor carrying the turbine wheel on a long, strong, and above all well-centred axis. The cooled axis thus passes through the insulating wall.

A thermostat may be compared to a control element. Thermostats that are to be used for gas chromatography must be able to make rapid adjustments to the temperature. The necessary temperature constancy must lie between ± 0·1°C and ± 0·5°C, depending on the height of the operating range.

As the column generally reacts within a few seconds to temperature fluctuations—the masses under consideration are fairly small—a considerably higher quality is required of the thermostat than would be expected from a required temperature constancy of only $\pm\ 0{\cdot}1°C$.

The temperature control system must have a low dead time (seconds) and a steep start up value.

Temperature control may be intermittent (on-off); this is the case for most laboratory thermostats. In this case the maximum temperature variation—from the on-off switching of the heating—must certainly be less than $\pm\ 0{\cdot}1°C$, and the switching period (heating-not heating) must be as much below 30 seconds as possible.

Temperature control may be continuous (electronic). Thermostats with on-off temperature control are only suitable if they have an anticipatory action as a means of improving the quality of the control. With this aid to control technology the intermittent controller has almost the properties of the continuous controller.

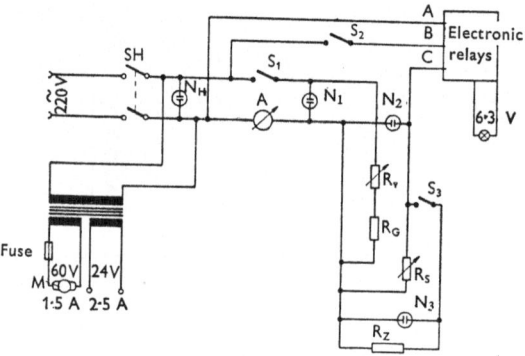

Figure 71. Example of a simple electrical layout for a gas thermostat.

S $_{\text{H-3}}$ = switches
R_A = basic heating (e.g. $160{-}180\Omega$; $1{\cdot}5$ A)
R_S = peak heating (e.g. 500Ω; $0{\cdot}5$ A)
R_Z = additional heating for initial heating (e.g. 70Ω; 3 A)
R_V = control member for the basic heating (e.g. 100Ω)
N $_{\text{H-3}}$ = neon lamps
M = motor for propeller

Normally for liquid thermostats (e.g. the Höppler type) an anticipatory action is used in conjunction with the controller, which switches the total heat output on and off. The switching frequency and also the temperature constancy of the thermostat can be increased considerably if only a part of the heat energy is switched on and off by the controller. This process is called the basic load-peak load switching system.

A gas thermostat with a correctly set basic heating and a peak heating controlled by a contact thermometer with anticipatory action is practically as good as a continuous thermostat, and has a much simpler circuit. The continuous thermostat needs a great deal of electronic equipment; on the other hand it is more convenient to service. *Figure 71* is an example of a circuit for a gas thermostat.

A basic load-peak load thermostat has normally the following elements as control elements:

Adjustable part: electrical resistance or variable transformer
Servomotor: electronic relay
Controller: contact thermometer with anticipatory action

An electronic relay is to be preferred to a normal relay as a servomotor because it puts far less a load on the contact thermometer. This is particularly important at high switching rates; over a long period the contact-breaking sparks in particular cause disturbances due to metal oxide deposits on the surface of the mercury. *Figure 72* shows the circuit for the electronic relay.

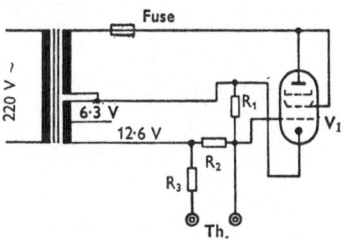

Figure 72. Diagram of the electronic relays;
V_1 = EL12 or EL84
R_1 = 300 kΩ
R_2 = 1 MΩ
R_3 = 10 kΩ
Th = contact thermometer

The anticipatory action is achieved by bringing a part of the heating coil for the peak heating into the immediate neighbourhood of the bulb of a contact thermometer. When the peak heating is switched on, the contact thermometer becomes hot more rapidly than would correspond to the true temperature of the surroundings. It is essential that the thermal inertia of the peak heating must be as small as possible. This may be achieved by the use of open heating spirals which are only clamped at a few points. In many cases the heating coil can be made from thin aluminium sheet (see *Figure 73*).

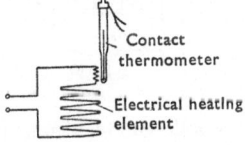

Contact
thermometer

Electrical heating
element

Figure 73. Principle of anticipatory action.

Basically the peak heating should have as little mass as possible. Ceramic heating bodies are thus unsuitable, and heavy metal constructions are equally to be avoided.

For the heating up of the gas thermostat the maximum possible heat output is switched on (about 1,200 W for a normal small thermostat which operates up to 300°C). For continuous operation of the same instrument with correct thermal insulation about 250 to 300 W are required, to compensate for the heat loss from the thermostat.

The air whose temperature is to be controlled can be given an external or an internal circulation. Both methods of construction have advantages and disadvantages. *Figures 74* and *75* show the principal arrangements.

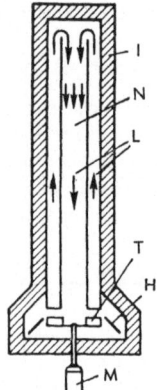

Figure 74. Diagram of a symmetrical air thermostat.

I = insulation; N = useful space; L = air circulation; T = turbine wheel; H = heating and temperature element for control of the heating; M = motor for the turbine wheel (should cause very strong air circulation, if possible several thousand litres per minute for a 10 l. thermostat.)

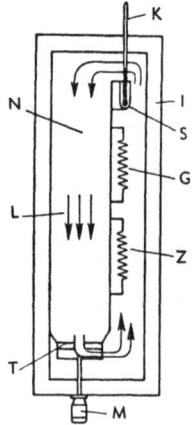

Figure 75. Diagram of a non-symmetrical air thermostat (greater useful space)
I = insulation; S = peak heating; G = basic heating; Z = additional heating; N = useful space; L = air circulation; T = turbine wheel; M = motor for the turbine wheel.

A continuous thermostat normally has the following elements as control elements:

Adjustable part:	electrical heating voltage
Servomotor:	vacuum tube
Controller:	resistance thermometer or thermistor.

The desired temperatures are obtained by adjustment of resistances (small potentiometers will do). Temperature programmes are also carried out by adjusting potentiometers with the aid of small synchronous motors.

For the control of detectors, switch valves or other small parts which need to be kept at a constant temperature, the so-called static electrical thermostats are used, which also use anticipatory action and solid contact thermometers. *Figure 76* shows the circuit together with the values needed for a

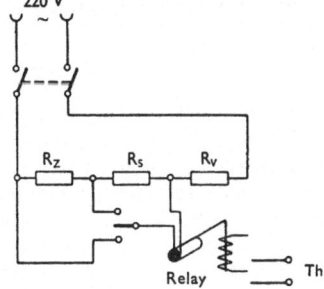

220 V

Figure 76. Diagram of a simple static thermostat.

R_z = additional heating (400 Ω)
R_s = peak heating (600Ω)
R_v = basic heating (1500 Ω)
Th = contact thermometer, bimetallic strips, etc.

R_z R_s R_v

Relay   Th.

thermal conductivity cell in continuous operation at about 150°C. The anticipatory action is again achieved by bringing part of the peak heating coil into the immediate neighbourhood of the bulb of a contact thermometer (or some other type of measuring element).

The proportion of the peak heating used for the anticipatory action depends primarily on the properties of the contact thermometer, and thus on its sensitivity in mm/°C scale length and on the mass of mercury. The output of the anticipatory action N_R should be as small as possible. Its relation to M and E is approximately:

$$N_R \sim \frac{M}{E} \cdot k$$

where M is the mass of the mercury, E is the sensitivity in mm mercury scale/°C and k = dimensional constant.

The thermal inertia of the anticipatory action must in any case be less than the thermal inertia of the contact thermometer.

Further details on the construction of a good gas thermostat are given by, among others, Ashbury, Davies and Drinkwater[8], and Keulemans[13].

Heat Insulation

The insulation, which of necessity must be constant, will only enable the thermostat to give a satisfactory performance over a wide temperature range if the heat loss can be artificially changed.

At lower temperatures the heat loss may be increased by cooling or by removal of part of the hot temperature control air. For high temperatures the insulation must be arranged so as to ensure good operation over a wide range. The heat loss of the thermostat should thus remain practically constant and independent of the operating temperature, and then optimum operating conditions may be expected over a wide temperature range. (Thus good insulation is necessary.)

For operating temperatures up to 250°C a tightly packed 50 mm layer of glass wool is sufficient for thermal insulation. At an internal temperature of 200°C the outer wall temperature of a thermostat insulated with a 40 mm layer of diatomaceous earth corresponds to about 60 to 70°C. The insulating power of foam glass* is so great that under the same conditions the external temperature would be 25 to 30°C.

Variable Temperatures

A truly satisfactory linear heating of a gas chromatograph with a rate of heating of e.g. 1°/min cannot be achieved without programmed temperature control. Even slight changes in heat conduction cause errors in the overall result so great that for non-isothermal chromatography the use of electronic equipment is an unavoidable necessity.

Harrison, Knight and Heath[11] describe a temperature programme controller with which they were able to get good results even in the more

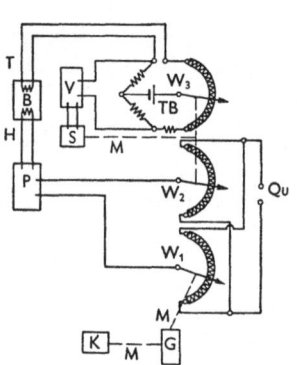

Figure 77. Diagram of a programmed temperature control system for chromathermography.
K = synchronous motor
G = three speed drive
M = mechanical connections
W_1 = 100Ω precision resistance
$W_{2,3}$ = mechanically coupled 100Ω precision resistances.
P = proportional relay
B = oil bath
H = heating
T = resistance thermometer
TB = temperature measurement bridge
V = amplifier
S = servomotor
Qu = source of voltage (see ref. 11).

difficult working range of 20–100°C. The block circuit diagram is given in *Figure 77*. The authors use an oil bath for lower temperatures up to 100°C.

By means of a synchronous motor the voltage is increased linearly with time from 0 to 10 v. The programming arrangement given by Harrison *et al.*[11], provides for three different rates of temperature rise. It is of course possible to carry out any desired temperature programme by means of programme pulleys or similar known measures.

A temperature measuring element—in the example chosen a resistance thermometer—is included as an arm of a Wheatstone bridge. For the second arm a precision potentiometer is inserted, the axis of which is connected to

* A new insulating material, analogous to foamed plastics.—Translator.

a servomotor. The linear or programmed increase in the voltage produced by a synchronous motor across the potentiometer W_1 is, by means of the temperature measurement bridge, compared with the voltage increase produced by the resistance thermometer; the voltage difference, which may be regarded as a measure of the delay or advance in the temperature of the oil bath, can be amplified and passed on to a servomotor. This adjusts the heating output of the temperature control system. The block circuit diagram shown in *Figure 77* shows a programmed temperature control system which transfers the difference in voltage between the programme voltage source and the temperature measuring element to a proportional relay which controls the heating current for the temperature bath. The authors concerned used two En 32 thyratrons and an EF 86 amplifier tube as switch elements in the proportional relay.

For less precise operation, temperature programming can be carried out as follows:

A synchronous motor works a wire spool of sufficient strength which, with the aid of a fine steel wire, turns a programme pulley on the axis of a variable transformer. The rotation of the variable transformer axis causes the heating voltage at the transformer output to rise. The temperature of the part of the instrument concerned rises approximately linearly with time if

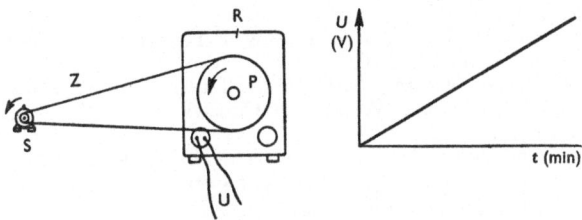

Figure 78. Simple programmed temperature control.
S=synchronous motor; Z=driving belt; R=regulating transformer; P = programme pulley; U = controlled voltage.

the transformer voltage is increased linearly. Irregularities are balanced by means of a programme pulley. The form of this is obtained by first using an exactly round pulley and obtaining the corresponding temperature-time curve. From the irregularities in the latter the form of the programme pulley on the variable transformer is calculated (see *Figure 78*).

Naturally any slight alteration in the insulation conditions and thermal capacities will upset the linearity of the temperature-time curve thus obtained.

References

1. AMBROSE, D., KEULEMANS, A. I. M. and PURNELL, J. H., *Analyt. Chem.*, 1958, **30**, 1582.
2. GREENE, S. A., MOBERG, M. L. and WILSON, E. M., *Analyt. Chem.*, 1956, **28**, 1369.

3. EVANS, J. B. and WILLARD, J. E., *J. Amer. Chem. Soc.*, 1956, **78,** 2908.
4. GRIFFITHS, J., JAMES, D. and PHILLIPS, C., *Analyst*, 1952, **77,** 897.
5. BERRIDGE, N. J. and WATTS, J. D., *J. Sci. Fd. Agric.*, 1954, **5,** 417.
6. KÖGLER, H., *Chem. Tech., Berlin*, 1957, **9,** 400.
7. KAMER, J. H. VAN DE, GERRITSMA, K. W. and WANSINK, E. J., *Biochem. J.*, 1955, **61,** 174.
8. ASHBURY, G. K., DAVIES, A. J. and DRINKWATER, J. W., *Analyt. Chem.*, 1957, **29,** 918.
9. LICHTENFELS, D. H., FLECK, S. A. and BUROW, H. F., *Analyt. Chem.*, 1955, **27,** 1510.
10. PATTON, H. W., LEWIS, J. S. and KAYE, W. I., *Analyt. Chem.*, 1955, **27,** 170.
11. HARRISON, G. F., KNIGHT, P., KELLY, R. P. and HEATH, M. T., *Gas Chromatography*, ed. D. H. Desty, Butterworths, London, 1958, p. 216.
12. WET, W. J. DE and PRETORIUS, V., *Analyt. Chem.*, 1958, **30,** 325.
13. KEULEMANS, A. I. M., *Gas Chromatography*, Reinhold Publ. Corp., New York, 1957.

Further Literature on Problems of Thermo Gas Chromatography

DREW, C. M. and McNESBY, J. R., *Vapour Phase Chromatography*, ed. D. H. Desty, Butterworths, London, 1957, p. 213.
FREDERICKS, E. M. and BROOKS, F. R., *Analyt. Chem.*, 1956, **28,** 297.
GREEN, S. W., *Vapour Phase Chromatography*, ed. D. H. Desty, Butterworths, London, 1957, p. 388.
LICHTENFELS, D. H., FLECK, S. A., BUROW, F. H. and COGGESHALL, N. D., *Analyt. Chem.*, 1956, **28,** 1376.
NOGARE, S. D. and BENNETT, C. E., *Analyt. Chem.*, 1958, **30,** 1157.
NUNEZ, L. J., ARMSTRONG, W. H. and COGSWELL, H. W., *Analyt. Chem.*, 1957, **29,** 1164.
RYCE, S. A. and BRYCE, W. A., *Analyt. Chem.*, 1957, **29,** 925.
SIMMONS, M. C. and SNYDER, L. R., *Analyt. Chem.*, 1958, **30,** 32.
ZLATKIS, A., *Analyt. Chem.*, 1958, **30,** 332.

2.7. INDUSTRIAL GAS CHROMATOGRAPHS AND OTHER SPECIAL APPARATUS

The following section deals with special arrangements of the individual components previously described.

Analytical Apparatus for High Boiling Liquids

High Temperature Instruments

Carrier gas: Helium; in many cases hydrogen, argon or nitrogen.
Sample injector: Microdipper, pneumatically controlled by-pass system.
Column: Short packed column, 4–6 mm diameter.
Detector: High temperature thermal conductivity cell for limited use; micro flame ionization detector to be preferred; combustion of substance and estimation of CO_2 in helium or conversion of the water of combustion to hydrogen where helium, argon or nitrogen is used as carrier gas are all possible; in these cases normal thermal conductivity cells may be used.

Reports on high temperature instruments are given by Cropper and Heywood[1] for work up to 200°C, Dijkstra et al.[2] and Beerthuis et al.[3] for work up to 260°C, and Hawkes [4] and Ashbury et al.[5] for work up to 300°C.

A high temperature instrument working up to 500°C is described by Wachi, F. M. (Ph.D. Thesis, University of Illinois, 1959, *Dissertation Abstr.*, 1959, **20**, 53).

Operation at 450°C is described by Baxter et al.[27].

More detailed data, in particular on the use of negative pressures at such temperatures, are given by Adlard et al.[6] for separations of paraffins up to C_{40} and fatty acids up to C_{36}.

Further references on apparatus for work at temperatures up to 450°C will be found in papers by Lewis and Patton[7], and Taylor[8].

The literature on gas chromatography at high and very high temperatures is very much on the increase. In principle the instruments are very similar. Further references will be found at the end of the list of numbered references.

In the construction and operation of such instruments the following points should be noted.

All rotating parts must, of course, have their axle mounting in the cold region of the apparatus. Only at great expense is it possible to run sliding bearings satisfactorily over a long period at temperatures above 100°C.

Very good heat insulation is especially important. The amount of heat lost through metallic tubes connecting the hot and cold parts of the instrument is generally underestimated. It is recommended to use the so-called flash heaters at such transition points (see under microdippers). Especial care must be taken in the choice of packing material for the connecting points of the individual parts of the apparatus. Suitable materials are mild aluminium, kautasit (with reservations), asbestos thread-powdered teflon mixture, and europin. All materials should have coefficients of expansion matched to one another.

Extreme difficulties occur if the carrier gas is not kept completely free from oxygen, because even traces of this substance will bring about the oxidative decomposition of the already strongly loaded liquid phase. Since the vapour pressure of the liquid phase is generally an unsettling factor it is recommended that high temperature instruments should be completely symmetrical in construction. No matter what type of detector is used, the apparatus should have a reference section and a measuring section, the reference section having a column similar to that in the measuring section (see *Figure 79*).

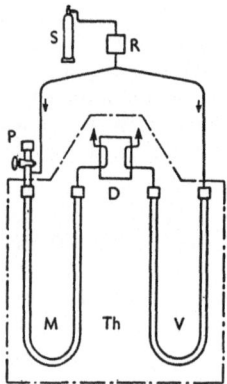

Figure 79. Completely symmetrical apparatus for very high temperatures.

S = source of carrier gas completely free of oxygen; R = gas flow rate controller (one for each branch of the gas path); P = sample injector; M = measuring column; V = comparison column; D = detector with measuring and control sides; Th = thermostat.

The use of the thermal conductivity cell as a direct detector in high temperature instruments is not to be recommended, since the measuring effect itself is considerably decreased by the rise in temperature. Thermal conductivity cells in conjunction with combustion tubes are, however, very suitable, for the combustion tube increases the sensitivity of detection simply by the stoichiometrical chemical process—from one mole of a C_{30} alkane 30 moles of CO_2 or 31 moles of hydrogen are formed. In this way the measuring cells can be operated at room temperature; quantitative analysis can be carried out with greater certainty, and even C/H ratios can be determined directly.

There are three possible variants of the combustion tube:

1. The substance is burnt over very finely powdered copper, cobalt or nickel oxide. The oxide is deposited on coarsely granular kieselguhr or alumina. The tube consists of VA steel or porcelain, is about 150 to 200 mm long and is run at 730 to 760°C. Its diameter should be 6–10 mm. To the tube is connected an H_2O absorber, for which it is preferable to use P_2O_5 or $Mg(ClO_4)_2$ (see *Figure 18*).

All connections should have the smallest possible dead volume.

2. The substance is burnt. The water formed is passed, together with CO_2, over iron turnings at a temperature of 800°C. VA steel is again suitable as tube material. The water is decomposed to release hydrogen. Finally the carbon dioxide is absorbed over asbestos impregnated with caustic soda, and the hydrogen is measured in a thermal conductivity cell.

3. The CO_2 and H_2 are separated over a short length of an appropriate chromatographic column.

For investigations where the C/H ratio is to be determined, the most suitable instruments are those operating chromathermographically with linear temperature increase. Otherwise the optimum separation and determination of the CO_2 and H_2 peaks is made very difficult.

Further data on combustion processes will be found in papers by, among others, Norem[9] and Green [10].

Multi-Column Instruments

The carrier gas, the sample injector and the detector used depend on the separation to be carried out.

The type of packing used for the columns, their length, the order of their arrangement and the temperatures used where necessary for a column also depend on the particular separation.

There are two types of multi-column instruments:

1. Instruments where the columns are connected in series. They may be used for special separations. They require only one detector and recorder each.

2. Instruments where the columns are connected in parallel. They may be used for the analysis of complex mixtures or for the separation of mixtures with very wide boiling point ranges. Often they can be used to shorten the time required for an analysis. Generally speaking, for one instrument several detectors and recorders are required.

As an especially simple case, let us consider parallel columns connected by a T-piece to a common sample injector and by another T-piece to the detector. Different flow rates in the two columns can be obtained by inserting different capillaries in the T-piece after the sample injector. With such columns it is possible to carry out, e.g. the analysis of He, N_2, O_2, H_2, CO, CO_2 and CH_4 together with higher hydrocarbons, using molecular sieve 5A or 13X in one column and a hydrocarbon packing in the other. The lengths of the two columns must, of course, be matched to one another. Given substances are recorded twice (once as a sum, once separately) (Brenner and Cieplinski[26]).

While the apparatus required for the first type of instrument differs only slightly from that normally used, special constructional measures are required for instruments where the columns are connected for part of the time in series and part of the time in parallel.

A simple case is that where the mixture contains both high and low boiling components, i.e. when the boiling points of the components differ greatly. *Figure 80* shows in outline the type of instrument used for such a task. The

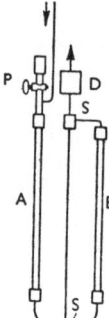

Figure 80. Two stage instrument with pneumatically controlled diaphragm valves (see also Figure *82*). P = sample injector; A = column A; B = column B; S = switch valve of type shown in *Figure 82*; D = detector.

substance first passes through column A. The rapidly travelling components then pass through column B and are recorded by the detector. In the meantime the more slowly moving components have reached the end of column A and now by turning the two-way tap are passed directly into the detector via a short capillary tube. The dimensions of the capillary are such that it produces the same pressure drop as column B would produce. In this way no pressure differences and variations in the gas flow rate occur on the switchover.

There are three different ways of operating such an instrument:

1. Columns A and B have the same packing and temperature. The instrument serves merely to shorten the time needed for the analysis of mixtures which contain both fast and slow moving components.

2. Column A has the same packing as column B but is at a higher temperature. In principle the result is the same as for 1 but with considerably greater saving of time; or it can be used to carry out an analysis of a mixture with an extremely unfavourable combination of high and low boiling compounds without preliminary separation.

3. The properties of column A are different from those of column B. With this arrangement complicated separations can be carried out which would not be possible on a simple instrument.

The essential point is that column B is shut down as soon as A is directly coupled to the detector.

An instrument of particularly simple construction was described by Green[13]; it is shown in *Figure 81*, but possesses a number of defects.

It is expedient to construct the two way tap, which is shown in *Figure 80* between columns A and B, as a small diaphragm valve controlled by compressed air, as shown in cross section in *Figure 82*. Switching can then be carried out without disturbing the apparatus; no questions of lubricants, gas tightness, temperature sensitivity etc. arise.

Especial care must be taken joining the connecting elements. Dead space must be very carefully avoided. Where necessary this may be filled in with glass rods or spheres.

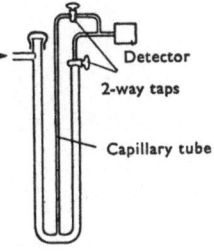

Figure 81. Simplest type of two stage instrument.

Detector

2-way taps

Capillary tube

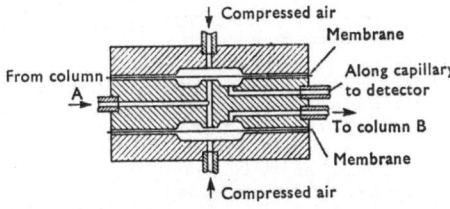

Compressed air

Membrane

From column A

Along capillary to detector

To column B

Membrane

Compressed air

Figure 82. Cross section of diaphragm valve based on Tschako tap. Steel diaphragm with conical seal or silicone rubber.

Columns Connected in Parallel

Many complicated separations can be performed only with the use of columns connected in parallel.

The plan for an apparatus as shown in *Figure 13* in the section on columns differs from the flow sheet in *Figure 83* in that, in the latter case, the number of needle valves required is equal to the number of columns run in parallel. It is, of course, necessary to have an appropriate number of recorders available with the detectors. Further details on the apparatus will be found in the work of Simmons and Snyder[14]. The essential thing is that for columns run in parallel the gas flow resistance should not alter on switching from the series to parallel arrangement. This may be managed by the insertion of fine needle valves which possess no dead space, or by the use of capillaries. Compressed air controlled small diaphragm valves are once again used as branch and switch elements. There are several possible gas path arrangements which can be used. The author recommends the arrangement shown in *Figure 83*, which differs from the arrangements given in the literature in two essential points. No flow resistance is inserted into the gas stream carrying the substance; the use of such a device is often a source of trouble. The gas path is extremely simple. *Figure 84* shows the mechanical arrangement of the gas path switch that is used.

The only types of detector that can be used with the arrangement shown in *Figure 83* are those which are gastight and in which the substance is not decomposed. Especially suitable are thermal conductivity cells and gas density meters.

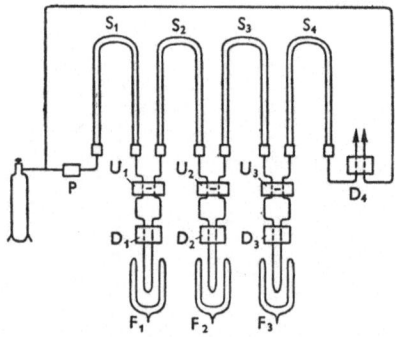

Figure 83. Gas path arrangement for a four stage instrument. Gas flows from cylinder through sample injector P into column S_1 (e.g. at 200°C) through the measuring side of detector U_1 and, by-passing cooling trap F_1 by the short route of the diaphragm valve in *Figure 84,* i.e. from A to B in *Figure 84,* directly into column S_2 (e.g. at 150°C), again through the measuring side of detector U_2 and the valve directly into column S_3, etc. If the distance between the peaks registered by detector D_4 is large enough, valve D_3 is first opened, then in turn, valve D_2, etc. The comparison gas flows from the cylinder through detectors D_4, U_3, U_2, U_1 (not apparent from figure).

For all columns a continuous stream of carrier gas is used. After the substance has been recorded by the individual detectors it is either frozen out of the gas stream or else passed into the neighbouring column. Switching does

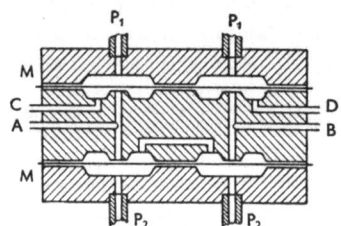

Figure 84. Cross section of gas tap needed for instrument in *Figure 83.*

M = diaphragm
A = from detector U_1
B = to column S_2
C = to cooling trap F_1
D = from cooling trap F_1 to column S_2
$P_{1,2}$ = compressed air.

not produce any pressure fluctuations. With the arrangement shown it is easy to notice if the freezing out is insufficient. The time at which it is necessary to switch over can be read off from the individual detectors. Inert gases do not cause any trouble, as they have, where necessary, to be passed into the last column. Detailed data on freezing out will be found in the next section.

Apparatus for Preparative Work

Preparative instruments are used for the production of pure individual compounds and for enriching a compound which is of interest but which is present only in trace amounts.

Such instruments are particularly simple when they are used for the extraction of gaseous substances which are insoluble in caustic potash. High purity air-free carbon dioxide is used as carrier gas and the fractions of interest are passed into an azotometer. Such an instrument is built up from the same components as a normal Janak analytical instrument; however the dimensions of the column are longer. Data on the dimensions necessary are found in the section on columns.

Instruments for the preparation of medium and high boiling substances are quite complicated.

It is expedient to keep the column performance high. Naturally this limits the amount of mixture that can be added per working cycle to a few grams (1–5 g per cycle at 30 mm diameter, 10–70 g per cycle at 75 mm diameter).

In order to obtain a worthwhile output of product, the process must be repeated as often as possible. This leads finally to the use of programmed automatic instruments. Their output, however, for a reasonable expenditure on construction and materials, still only lies in the range of a few hundred grams per day.

As well as possessing a good column performance, an instrument which is used in continuous operation needs mechanical and electrical switch elements which work extremely well. But the problem of obtaining the product to be frozen out without any loss becomes difficult when the substance concerned tends towards aerosol formation. Naturally there are ways of overcoming this property. But it is obvious that the cost of the apparatus will be quite high.

Preparative instruments have been described, among others, by Whitham[15], Atkinson and Tuey[16] and Kirkland[17].

In addition to this, de Wet and Pretorius[29] have concerned themselves with the theory of preparative gas chromatography. Data about the limits of preparative gas chromatography will be found in the works of, among others, Bens and McBride[30].

Bayer, Hupe and Witsch[34] have described a preparative gas chromatograph in which up to 100 ml of a mixture can be separated per operating cycle. By means of programmed repetition of the process, up to 5 l. per day of suitable vaporizable substances can be separated (using a cycle of less than 10 min).

In preparative gas chromatography the liquid phase has a special influence on the purity of the fractions obtained. The use of high temperatures can become critical for this, since if the liquid phase has a reasonably large vapour pressure it will then vaporize in considerable quantities and cause a marked contamination of the product obtained. Kovats recommends the following liquids for preparative gas chromatography:

non-polar:	silicone oil DC 710	200–350°C
semi-polar:	Apiezon L	200–300°C
polar:	Emulphor O BASF	200–270°C

The column packing must be freed from volatile matter prior to use by heating to 350°C for a short time with large quantities of completely oxygen-free gases (N_2). This is best carried out in a fluidized bed process; a short 100 mm wide tube can be used as the fluidized bed.

The column packing should be kept in an airtight container, so that no trace of oxygen can dissolve in it.

The carrier gas for preparative gas chromatography must be completely free from oxygen. (The best arrangement is to circulate the gas as in the Megachrom manufactured by Messrs Beckman GmbH, Munich.)

Operating Conditions

Carrier gas:	Any.
Quantity of gas:	Up to 300 l./hr, but see the section on carrier gases.
Column:	Up to 30 mm in diameter, in special cases up to 200 mm with internal direction elements; up to 20 m in length, but see the section on columns.
Detector:	Thermal conductivity cell or flame ionization detector placed in a side stream (1/100).
Sample injector:	Preferably piston injection. To avoid the slight but very troublesome continual influx, the outlet connecting the pump to the column must be sealed off with additional valves. Special care must be taken over the complete evaporation of the substance. The rule that instantaneous point evaporation is to be recommended does not apply here. The pressure waves produced by the considerably greater quantities of substance should be kept as small as possible. The time needed for complete evaporation may be 10 seconds. Of particular importance is a sufficiently high evaporation temperature. It is a good idea to evaporate the substance completely before injecting it into the column.

Sample Collection Systems

The purity and yield of the product depend to a large extent on the method of operation of the sample collection system. This is the essential part of the preparative apparatus, and for this reason is described in more detail below.

A number of conditions must be fulfilled:

In order to freeze the relatively small amount of substance out of the large excess of carrier gas, the gas stream must, of necessity, be subjected to turbulent motion in the cooling trap. A laminar flow reduces the yield considerably. The mist that is formed with many substances must almost always be precipitated electrostatically. A means for doing this must therefore be provided if the instrument is going to be used for general purposes. It must be possible to switch the gas stream from the column to a large number of cooling traps by means of a temperature controlled and completely gastight switch system. Because of the large temperature drop between the heated connections, which are usually metal, and the cooling traps, the only

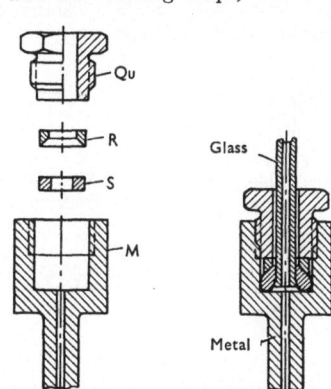

Figure 85. Fehlauer (Schkopau) screw pinch connection.

Qu = screw collar; R = bevel ring; S = silicone rubber or Teflon asbestos ring; M = metal tube with counter-ring.

materials which can be used are glass or quartz. The connection between the instrument and the cooling trap must be gastight. A ground ball and socket joint, or possibly a cone, is suitable, but the silicone rubber pinch screw arrangement is better (see *Figure 85*). Cooling traps of the types shown in

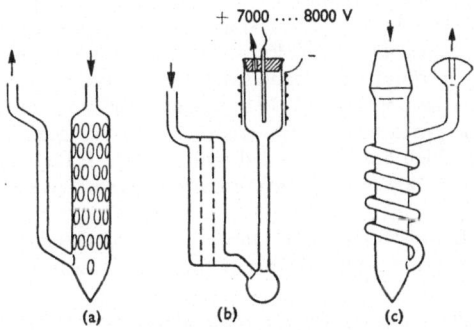

Figure 86. Cooling traps.

Figure 86 a and b are attached with the silicone rubber pinch screw arrangement; the rather wider tube has irregular constrictions (indentations) which give rise to turbulent flow. The frozen out substance may readily be collected in the drawn out tip at the lower end of the cooling trap. The form shown in *Figure 86*a has, together with the rather more complicated form shown in *Figure 86*c, the advantage that the thoroughness of the freezing-out process can be controlled by means of a thermal conductivity cell. *Figure 87* shows the arrangement that must be used for the gas path in this case. The carrier gas and substance issuing from the column are passed through the measurement side of the thermal conductivity cell and enter the cooling trap. The carrier gas freed from substance is then passed through the reference section of the thermal conductivity cell. From the base line recorded it can be seen whether the substance is being removed partially or completely. (In the latter case peaks appear beneath the base line.)

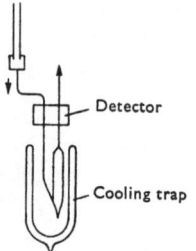

Detector

Cooling trap

Figure 87. Gas path arrangement for the control of freezing out (only for use with detectors that have measuring and reference sections).

The cooling trap shown in *Figure 86*b enables the application of high tension fields for the removal of stable mists. The internal electrode may consist of a steel rod (cathode) or a platinum wire; the external electrode applied to the glass tube can consist of a piece of thick wire gauze of brass or copper. Atkinson and Tuey[16] report that with a voltage of 7,000–8,000 V they were able to remove even the most obstinate mists satisfactorily. In their case they set the cathode as a steel tube within the cooling trap and used the central rod as anode. Both poles were kept firmly in position by means of a plastic stopper. Liquid nitrogen is only suitable for use as cooling medium when the carrier gas is free from oxygen, which could otherwise be precipitated. If the vapour pressures of the substances which are being prepared are fairly low then the cooling medium can be an organic liquid which has been cooled with dry ice. If the vapour pressures are very high it is best to introduce intensely cooled active carbon into the gas path.

Wehrli and Kovats[28] describe an effective centrifugal cooler for the separation of ice or cloud-forming substances, thus doing away with the need for the electrostatic process.

The switching elements can be the compressed air controlled small diaphragm valves which have already been mentioned.

Atkinson and Tuey[16] describe an 8-way tap controlled by an electric motor. The tap consists of a steel block with 8 borings and a rotating teflon disc with transverse channels. *Figure 88* shows the principle of such a tap. The author has also tested such devices. They all have the disadvantage that they cannot remain gastight when exposed to large variations in temperature. Rotary switches require an especially precise working of the sealing surfaces and a very good teflon material. The elegant diaphragm valves are to be preferred. However, high precision steel surfaces do enable use without sealing material and lubricants, as they can operate dry.

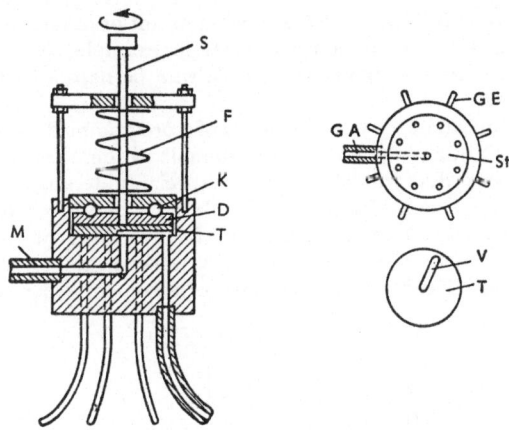

Figure 88. Rotating multiway tap (Atkinson and Tuey[16]). S = driving shaft; F = spring; K = ball bearing; D = pressure disc; T = disc of Teflon or other suitable material; M = central tube; GA = tube M = from column; GE = to cooling traps; V = radial groove in Teflon disc.

The magnetic valves needed for the control of diaphragm valves are available commercially. Napier and Rodda[18] give details of a multiple fraction collector which, however, is only of limited application. It uses glass taps set into metal blocks with synthetic resins, and is intended for use *in vacuo*.

Fully Automatic Instruments

For continuous process supervision, and in particular for process control, fully automatic instruments have been developed.

The vital part of the fully automatic instrument is the automatic sample injector. Of all the devices known at present the most suitable for the analysis of gases and liquids boiling up to 200°C is the diaphragm valve system of Hooimeijer *et al.* which was described in the section on sample injectors. Its method of operation depends once again primarily on the quality of the mechanical finish and on the quality of the switch diaphragms. Again, only the isothermal means of operation appears advisable for automatic instruments, since the take-off of the control signal from the instrument must be time-controlled. Automatic control over long periods of time can only be carried out with absolutely constant operating temperatures and a constant gas flow rate.

Efficient magnetic valves are a pre-requisite for the satisfactory functioning of the sample injector. Finally a versatile time programme controller will permit everything which may be required of an automatic system.

The production engineer naturally wishes to be able to read off the values directly without the need for further calculation; further, he does not want to be bothered with values which are unimportant or of no interest. In the analysis of a mixture of n components he may, for example, be concerned only with the recording of components 3 and 6, with perhaps a control impulse from component 2.

This problem can be solved with the aid of a programmed time controller. For this G.P.O. relays are particularly suitable as the chief constructional unit. The relevant data will be found in the specialist literature (on measurement and control technology). Detailed data are also given by Fischer[19], who described an automatic instrument for five different measurement circuits. The time controller described there works on the principle of condenser discharge. The plan for an automatic instrument is given in *Figure 89*.

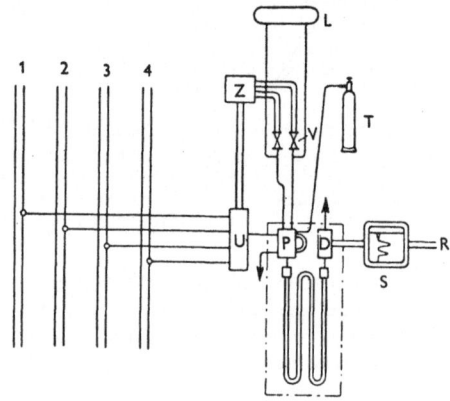

Figure 89. Diagram of an automatic apparatus for the supervision of several product lines.

Z = time controller and programmer; L = compressed air; T = carrier gas; V = magnetic valve; U = tap as shown in *Figure 88*; P = sample injector; R = control impulse; D = detector; S = recorder and evaluator.

By means of the switch U—for this both magnetic valves and piston pumps are only of limited application, but the multiple tap shown in *Figure 88* is very suitable—any product stream can be connected to the sample injector P for a given time.

By means of the time controller Z both the switch U and the sample injector P can be connected with and disconnected from the valve V and the recorder S according to a prearranged time programme.

If, for example, the switch U is set for the product stream 3, then product 3 will be flowing through the sample injection loop; the sample injector is being prepared for the analyis of 3 while the recorder is still registering the results for 2. At the end of the analysis of 2 the time controller switches the sample injection valve.

The product from stream 3 is now led past the sample injector; at the same time the switch U can be set to stream 4, so that in the mean time the connecting tubing can be washed out. The substance 3 in the sample injection loop is injected into the column. If, for example, the analysis of the first two components of mixture 3 is not of interest, the time controller switches off the paper transport mechanism of the recorder together with the deflection of the recorder carriage.

After 180 seconds the third and fourth components of substance 3 may appear; the time controller therefore switches the carriage mechanism of the recorder back on again, and between the appearance of the third and fourth components the paper transport is also switched on again for a short time. In the mean time the sample injector valve has been switched to a position where it can start preparing for the analysis of substance 4. It will be realized that only a simple switch arrangement will be needed to enable the apparently complicated electrical and mechanical switch mechanism to be carried out according to a time programme.

Further data on process control instruments are given by Wall[31], Wall and Baker[32] and Fourroux[33]. See also the literature references.

In the last few years, experience has shown that process control instruments must have two main types of property: they must first be quite specifically and optimally suited to the control problem, and secondly have exceptionally good working parts. A time programme controller which, if possible, can be influenced by the analytical result (in order that it may signal unforeseen deviations from the normal course), an exceptionally constant sample injector, and a good thermostat with a stable detector are the minimum requirements for an industrial instrument. These requirements sound obvious and yet are very difficult to fulfil, especially when it is considered that the instrument must work just as well after 3,000 hours as after 3 hours.

It is only very rarely that a suitable column can be obtained from stock, and so for this reason and for the reasons given above it is common to find home-made instruments, in both large and small works, although in the latter case they are of the 'string-and-scalingwax' variety.

Close and fruitful collaboration between the instrument manufacturers and the chemical industry is, however, giving rise to the production of commercially produced process control instruments of high quality, which will enable gas chromatography to take its rightful first place in process supervision and later in process control.

The firms Beckman Instruments USA (Beckman Instruments GmbH, München 45, Frankfurter Ring 115—see *Figure 90*) and Perkin Elmer Norwalk USA (Perkin Elmer & Co, Bodenseewerk, Überlingen am Bodensee—Process Vapour Fraktometer Model 184) have had many years experience in supplying industrial instruments.

Recently W. G. Pye & Co, Granta Works, Cambridge, England, have been offering a 'Process Analyser' fitted with an ionization detector.

Figure 90. Gas chromatography process control device produced by Beckman Instruments USA, or Beckman GmbH Munich.

Process control instruments are also supplied by Carlo Erba, Milano, Via C. Imbonati 24.

Process control instruments record the analytical result either as a line diagram, for which only the components of interest are recorded as lines which then must be converted by linear percentages into concrete quantitative values, or as a point diagram. The point diagram is preferable to the line diagram, since the latter cannot be read at a glance. The point diagram (Runge[35]) is formed by the printing of the end value of the recorder deflection as a point (coloured or with a numerical value). Runge reports that this process can be applied without using a great deal of electricity wherever the instrument has been set up for line diagram recording. The point diagram gives a clearly visible record, as is known from temperature recorders with several measuring points. The point values may correspond both to the peak heights and to the peak areas. The values are, however, still only analogue values. Integrators with print-out are better. With these a qualitative value (retention time) is printed at the same time as the digitally calculated quantitative value. The first instruments in this difficult but, for future development, unavoidable field have been made by Hartman and Braun AG, Frankfurt am Main—West 13, Gräfstrassse 97.

Trace Analysis Instruments

The excellent performance of gas chromatographic columns on the one hand, and the outstanding sensitivity of detectors on the other soon led to the development of instruments suitable for the trace analysis of vapours in air or gases and also for impurities in individual compounds.

There are two methods of increasing the sensitivity of gas chromatographic methods to the value required for trace analysis:

by trace enrichment, or
by improving the sensitivity of detection.

With the aid of mercury, or by suction, the sample is passed over intensely cooled active carbon columns at about −70°C. Sealing fluids should not be used for gas transport. The active carbon column is then electrically heated (about 250 to 300°C). The desorbed substance is passed by means of an air-free stream of carbon dioxide into an azotometer, where the trace is kept in a concentrated form. From this azotometer vessel the trace substance is passed into a normal but highly sensitive gas chromatograph. In spite of the three interconnected processes, these analyses are still so accurate that the result can be given with an error less than ± 5 per cent relative.

Figure 91 shows an arrangement which can be used in apparatus for indirect trace analysis in air (enrichment).

Bennett et al.[20] carried out an investigation into the increase of detector sensitivity by electronic amplification, and the type of detector most suitable for the process. They found that the most favourable signal-noise ratio was obtained with thermal conductivity cells using thermistors as measuring elements, and were thus able to obtain an instrument which could analyse 1 to 200 parts per million.

Using a thermal conductivity cell of normal sensitivity and the enrichment method Holzhäuser[21] was able to increase the sensitivity of the trace analysis of hydrocarbons in synthesis gas and air from a maximum of 1 to 0·1 parts per million.

For the enrichment of traces of hydrocarbons in synthesis gas and air, active adsorbents are used. At low temperatures (in the region of $-70°C$) in solid CO_2-acetone mixtures, adsorption is complete. The traces are then desorbed at higher temperatures and finally separated chromatographically and identified. Phillip *et al.*[22] used a very similar method. These authors dealt in particular with the problems of sample preparation and on the spot sampling.

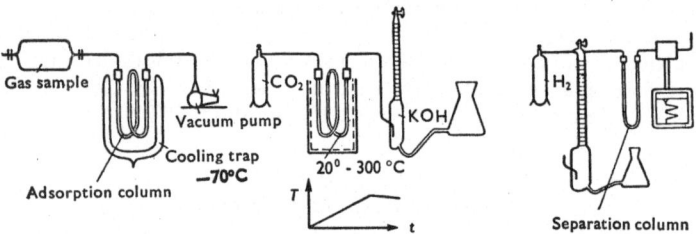

Figure 91. Diagram of a trace analysis instrument (e.g. for the determination of 10^{-6} per cent traces of hydrocarbons in air, synthesis gas, etc). Stage 1: enrichment over adsorption column at very low temperature; 2: thermal desorption and washing out into an azotometer using high purity CO_2 as carrier gas; 3: analysis of the enriched gas on a normal gas liquid chromatography column.

Bodnar and Mayeux[23] studied the determination of traces of lower alcohols, ethers and acetone in water with the aid of gas chromatographic methods. Boggus and Adams[24] used trace analysis as a purity (99 per cent) check in industrial process control.

Generally speaking, it is only the methods of gas chromatography that have enabled analytical investigations to be made in this otherwise very difficult field.

Trace analysis of polymerizable monomers is of particular importance, since the trace substances may have a positive or negative catalytic effect which can have a very strong influence on the properties of the polymer finally prepared. The tremendous savings that can be made in the search for oil and natural gas by trace analysis of the soil diffusion air (even from 2 m borings) to give a definite answer on the possibility of natural gas and oil reserves below the surface need only be mentioned in passing.

High sensitivity thermal conductivity cells enable traces at a concentration of 10^{-2} per cent to be recorded in direct analysis with certainty and traces at a concentration of 10^{-3} per cent to be recognized.

Ionization detectors can record up to 10^{-8} per cent (i.e. $0·0001$ p.p.m.) in direct analysis when set to maximum sensitivity. Water, which is generally present in large quantities, gives rise to particular trouble in such analyses. The use of liquid or solid drying agents during sample taking, injection or enrichment is not, however, to be recommended, since the mere contact with the drying agent will have a considerable effect on the composition of the trace and its absolute concentration. Further, in direct trace analysis the vapour pressure of the liquid phase can cause a considerable disturbance, and for this reason the greatest possible constancy of gas flow rate, temperature and pressure is required for the whole apparatus.

As the adsorption packings used in gas solid chromatography give a very high column performance and a complete symmetry of peaks (no more

Figure 92. Process control instrument produced by W. G. Pye and Co., Cambridge. Such instruments are fitted with high sensitivity β-ray ionization detectors and are thus in many cases even suitable for the automatic trace analysis of industrial products.

tailing) when loaded with only small quantities of substance, active column packings containing no liquid phase are to be preferred in direct trace analysis.

Further the preparation of model trace mixtures, in which the added concentration remains constant for some time, is almost impossible.

Surface adsorption by all the vessel walls and to an even greater extent by the lubricants used in the taps and joints gives rise to errors to powers of ten. The ageing of trace concentrations occurs with astounding rapidity.

These requirements must be borne in mind when trace analysis is to be carried out by the direct method. All the apparatus in the gas path must therefore be kept at the highest possible degree of cleanliness. It is also recommended that the liquid phase or active adsorbent used should be

159

specially chosen to suit the trace in question. A number of substances are absorbed until a saturation level is reached. It is, however, not possible to

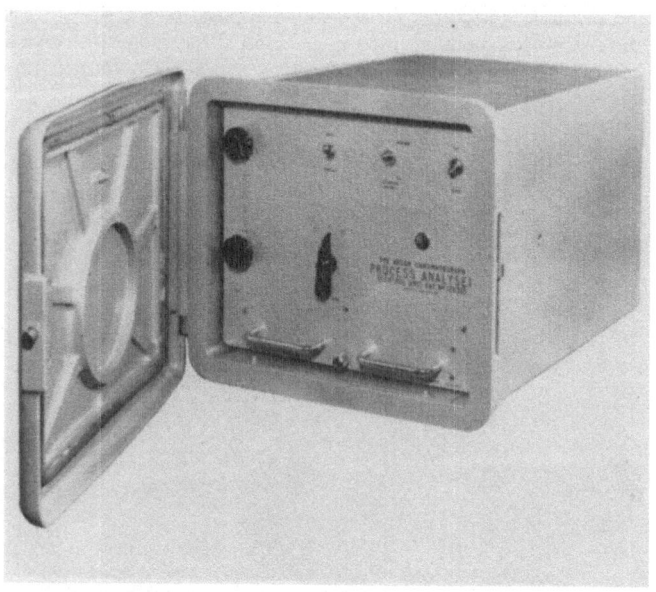

Figure 93. Control panel of a process control instrument for industrial gas chromatography (Pye).

give rules of general application here, since the optimum working conditions depend too much on the actual analysis being performed (see *Figures 92* and *93*).

Low Pressure Instruments

It was originally thought that if the whole column were run at reduced pressure the analytical times would be shortened just as if the temperature had been increased. However, this is now known to be incorrect, and low pressure instruments are now only used in exceptional cases. Prerequisites for the successful operation of low pressure instruments are a column with a flow resistance as low as possible, a completely gastight detector and a completely gastight sample injection system.

Only closed types of detectors, such as radiation ionization detectors or thermal conductivity cells, are suitable. As low pressure instruments are fundamentally more trouble-prone than high pressure instruments their use should be limited to those cases where it is absolutely necessary.

Carrier gas: nitrogen, argon or helium, as required.
Gas flow rate: about 1 l./hr for columns of 6 mm dia-
                                      meter.
Column length and diameter: should be chosen so that if possible the flow
                                      resistance causes a pressure drop of less
                                      than 0·2 atm.
Sample injector: preferably microdipper.

The maintenance of a constant gas flow rate causes a certain amount of difficulty. The author and his colleagues have been able to obtain a reasonably constant gas flow rate with the following arrangement. First a constant carrier gas pressure is obtained by means of a good diaphragm pressure controller. By means of a precision needle valve before the sample injector the first pressure reduction is produced. Between the detector outlet and the vacuum pump (oil pumps and water jet pumps are suitable) there is a second needle valve, which, however, must have a considerably greater aperture than the needle valve in front of the sample injector. The layout of a low pressure apparatus is shown in *Figure 94*.

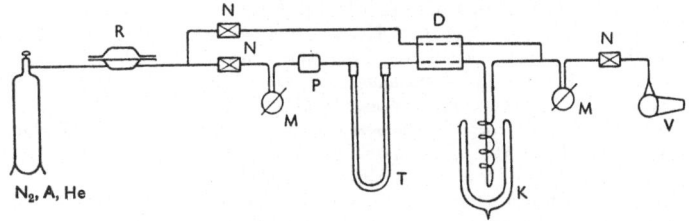

Figure 94. Diagram of a low pressure apparatus.
R = pressure controller; N = needle valve; M = manometer; P = sample injector; T = column; D = detector; K = cooling trap; V = vacuum pump

To avoid the contamination of the needle valve behind the detector which would otherwise certainly occur, a cooling trap which freezes all the components out of the carrier gas stream with liquid nitrogen is inserted. This is necessary, because otherwise also the oil in the vacuum pump or the performance of the water jet pump would be adversely affected.

Sorensen and Soltoft describe column packings with especially low pressure drop[25]. They impregnated very small steel turnings used for distillation column packings with about 1–3 wt. per cent of the liquid phase. Steel turnings 1·5 mm square in a column of 8 mm internal diameter at 200°C, 20 mm initial pressure and 0·5 l. N_2/hr produced a pressure drop of 4·5 mm/m. In comparison with this, a column packed with sodium chloride crystals under similar conditions gives a pressure drop of about 50 mm, and for columns packed with Sterchamol of grain size 0·3 to 0·5 mm it is about 200 to 300 mm.

Such columns have nevertheless a performance of 45 plates per metre. With a 4 m long column which has a pressure drop of only 16 mm and a performance of 180 plates it is just possible to separate the homologous series of the *n*-alkanes.

It must, however, be realized that it is only wise to use low pressure instruments when the pressure drop is very small; such columns possess only relatively small performances.

References

1. CROPPER, F. R. and HEYWOOD, A., *Nature, Lond.*, 1953, **172,** 1101; 1954, **174,** 1063.
2. DIJKSTRA, G., KEPPLER, J. G. and SCHOLS, J. A., *Rec. Trav. chim. Pays-Bas*, 1955, **74,** 805.
   idem, Vapour Phase Chromatography, ed. D. H. Desty, Butterworths, London, 1957, p. 222.
3. BEERTHUIS, R. K. and KEPPLER, J. G., *Nature, Lond.*, 1957, **179,** 731.
4. HAWKES, J. C., *Vapour Phase Chromatography*, ed. D. H. Desty, Butterworths, London, 1957, p. 266.
5. ASHBURY, G. K., DAVIES, A. J., and DRINKWATER, J. W., *Analyt. Chem.*, 1957, **29,** 918.
6. ADLARD, E. R. and WHITHAM, B. T., *Gas Chromatography*, ed. D. H. Desty, Butterworths, London, 1958, p. 351.
7. LEWIS, J. S. and PATTON, H. W., *Gas Chromatography*, ed. Coates, Noebels, Fagerson, Academic Press Inc., New York, 1958, p. 145.
8. TAYLOR, B. W., *Gas Chromatography*, ed. Coates, Noebels, Fagerson, Academic Press Inc., New York, 1958, p. 155.
9. NOREM, S. D., *Gas Chromatography*, ed. Coates, Noebels, Fagerson, Academic Press Inc., New York, 1958, p. 191.
10. GREEN, G. E., *Nature, Lond.*, 1957, **180,** 295.
11. HEFT, C., Diploma thesis for the Faculty of Maths and Natural Sciences, Karl-Marx University, Leipzig (1958).
12. BREDEL, H., diploma thesis for the faculty of Maths and Natural Sciences, Karl-Marx University, Leipzig (1958).
13. GREEN, S. W., *Vapour Phase Chromatography*, ed. D. H. Desty, Butterworths, London, 1957, p. 388.
14. SIMMONS, M. C. and SNYDER, L. R., *Analyt. Chem.*, 1958, **30,** 32.
15. WHITHAM, B. T., *Vapour Phase Chromatography*, ed. D. H. Desty, Butterworths, London, 1957, p. 194.
16. ATKINSON, E. P. and TUEY, G. A. P., *Gas Chromatography*, ed. D. H. Desty, Butterworths, London, 1958, p. 270.
17. KIRKLAND, J. J., *Gas Chromatography*, ed. Coates, Noebels, Fagerson, Academic Press Inc., New York, 1958, p. 203.
18. NAPIER, I. M. and RODDA, H. J., *Chem. & Ind.*, 1958, 1319.
19. FISCHER, J., *Gas-Chromatographie 1958*, ed. H. P. Angele, Akademie-Verlag, Berlin, 1959, 22.
20. BENNETT, C. E., NOGARE, S. D., SAFRANSKI, L. W. and LEWIS, C. D., *Analyt. Chem.*, 1958, **30,** 898.
21. HOLZHÄUSER, H., diploma thesis for the faculty of Maths and Natural Sciences, Karl-Marx University, Leipzig (1959).
22. WEST, P. W., SEN, B. and GIBSON, N. A., *Analyt. Chem.*, 1958, **30,** 1390.
23. BODNAR, S. J. and MAYEUX, S. J., *Analyt. Chem.*, **30,** 1958, 1384.
24. BOGGUS, J. D. and ADAMS, N. G., *Analyt. Chem.*, 1958, **30,** 1471.
25. SORENSEN, I. B. and SOLTOFT, P., *Acta chem. scand.*, 1956, **10,** 1673.

References

26. BRENNER, N. and CIEPLINSKI, E., *Ann. N.Y. Acad. Sci.*, 1959, **72**, 705.
27. BAXTER, R. A. and KEEN, R. T., *Analyt. Chem.*, 1959, **31**, 475.
28. WEHRLI, A. and KOVATS, E., *J. Chromatog.*, 1960, **3**, 313.
29. DE WET, W. J. and PRETORIUS, V., *S. Afr. industr. Chem.*, 1959, **13**, 105.
30. BENS, E. M. and McBRIDE, W. R., *Analyt. Chem.*, 1959, **31**, 1379.
31. WALL, R. F., *Industr. Engng Chem.*, 1959, **51**, April, 73A.
32. WALL, R. F. et al., *Ann. N.Y. Acad. Sci.*, 1959, **72**, 739.
33. FOURROUX, M. M., *Proc. 37th Ann. Conv. Nat. Gas. Ass. Amer.*, 1958, 16.
34. BAYER, E., HUPE, K. P. and WITSCH, H. G., Karlsruhe, Paper presented at Analytical Meeting of the Gesellschaft Deutscher Chemiker, Munich, Oct. 1960.
35. RUNGE, H., Ludwigshafen, private communication. Paper by H. Kienitz, Dechema Colloquium on Gas Chromatography, Frankfurt, Oct. 1960.

High Temperature Instruments

Up to 350°: NOGARE, S. D. and SAFRANSKI, L. W., *Analyt. Chem.*, 1958, **30**, 894.
Up to 400°: OGILVIE, J. L., SIMMONS, M. C. and HINDS, G. P. jr., *Analyt. Chem.*, 1958, **30**, 25.
Up to 500°: FELTON, H. R., *Gas Chromatography*, eds. Coates, Noebels, Fagerson, Academic Press Inc., New York, 1958, p. 131.

Cooling Traps and Fraction Dividers

BERRIDGE, N. J. and WATTS, J. D., *J. Sci. Fd Agric.*, 1954, **5**, 417.
BOGGUS, J. D. and ADAMS, N. G., *Analyt. Chem.*, 1958, **30**, 1471.
CALLEAR, A. B. and CVETANOVIC, R. J., *Canad. J. Chem.*, 1955, **33**, 1256.
EVANS, D. E. M. and TATLOW, J. C., *J. Chem. Soc.*, 1955, 1184.
KEPPLER, J. G., SCHOLS, J. A. and DIJKSTRA, G., *Rec. Trav. chim. Pays-Bas*, 1956, **75**, 965.
WEINSTEIN, A., *Analyt. Chem.*, 1957, **29**, 1899.

Automatic Instruments for Control Purposes and with a Number of Columns

AYERS, B. O., *Gas Chromatography*, eds. Coates, Noebels, Fagerson, Academic Press Inc., New York, 1958, p. 249.
CLAUDY, H. N., HELMS, C. C., SCHOLLY, P. R. and BRESKY, D. R., *Ann. N.Y. Acad. Sci.*, 1959, **72**, 779.
GREENE, S. A., *Analyt. Chem.*, 1959, **31**, 480.
HURN, R. W., CHASE, J. O. and HUGHES, K. J., *Ann. N.Y. Acad. Sci.*, 1959, **72**, 675.

Automatic Instruments for Multi-Component Mixtures

HELMS, C. C. and CLAUDY, H. N., *Gas Chromatography*, eds. Coates, Noebels, Fagerson, Academic Press Inc., New York, 1958, p. 269.

Preparative Instruments

KOVATS, E., SIMON, W. and HEILBRONNER, E., *Helv. chim. acta*, 1958, **41**, 275.
POLLARD, F. H. and HARDY, C. J., *Chem. & Ind.*, 1956, 527.
STAHL, W. H. and LEVY, E. J., *Analyt. Chem.*, 1956, **28**, 1058.
SYKES, A., TATLOW, J. C. and THOMAS, C. R., *Chem. & Ind.*, 1955, 630.
TURNER, D. W., *Nature, Lond.*, 1958, **181**, 1265.

YUROVSKII, Y. M., Instrument for determination of toxic gases in mines, *Bezopastnost' Truda v Promyshlennost'*, 1957, **8,** 29.

ZHUKHOVITSKII, A. A., TURKEL'TAUB, N. M. and GEORGIEVSKAYA, T. V., Continuous chromathermography, *Dokl. Akad. Nauk SSSR*, 1953, **92,** No. 5, 987.

TURKEL'TAUB, N. M., Chromatographic titrimetric gas analyzer, *Zavodskaya Lab.*, 1949, **15,** 653.

TURKEL'TAUB, N. M. and ZHUKHOVITSKII, A. A., Continuous chromatography, *Geologiya Nefti.* 1957, **1,** 54.

TURKEL'TAUB, N. M., PORSHNEVA, N. V. and KANCHEEVA, O. A., A chromatographic thermochemical gas analyzer, *Zavodskaya Lab.*, 1956, **22,** No. 6, 735.

TURKEL'TAUB, N. M., Chromatographic method for separate determinations of micro-concentration of hydrocarbons in air, *Zhur. Anal. Khim.*, 1950, **5,** No. 4, 200.

3. THE ANALYTICAL RESULT

The following two sections describe processes for the qualitative and quantitative evaluation of chromatograms.

First a short account is given of the way in which a chromatogram must be interpreted before a qualitative evaluation can be undertaken. Methods and rules for the identification of unknown substances are given. Where possible, practical examples are given. Finally problems of quantitative analysis are discussed.

THE GAS chromatogram normally consists of a series of peaks when a differential detector is used for recording. If an integral detector (e.g. an automatic burette) is used a step-shaped integral curve is obtained. This is converted into a differential curve. From such a chromatogram the following may be read off at once:

1. The minimum number of components in the mixture (= number of peaks).

2. The major and minor components.

The following points must, however, be borne in mind:

With respect to 1:

(a) A peak may be composed of several substances without this being obvious. The peak width at half height $b_{\frac{1}{2}}$ has a value which may be calculated from the column performance n'

$$b_{\frac{1}{2}} \simeq t_{dri} \sqrt{\frac{5 \cdot 54}{L \cdot n}}$$
$$\simeq t_{dri} \cdot C$$

where C = constant

(b) The peak may be unsymmetrical or wider than corresponds to $\sim t_{dri} \cdot C$: in this case it certainly consists of more than one component. All

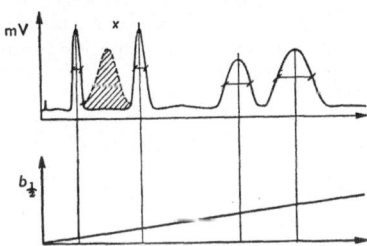

Figure 95. Relationship between the peak width at half height $(b_{\frac{1}{2}})$ and the retention time. Peak X in the upper chromatogram therefore clearly consists of several components.

we can tell from the number of peaks is the *minimum* number of components in the mixture. It is, for example, possible for the octane peak to contain all 18 isomers (see *Figure 95*).

165

(c) The chromatogram only gives information on the substances which can vaporize from the substance mixture within the time of the analysis. It is thus possible for substances to appear on the chromatogram considerably later, after the instrument has been switched off or after a new analysis has been started. It is also possible for considerable quantities of material to remain practically unvaporized under the conditions chosen.

(d) Reactions may occur with the column packing or the inner surface of the apparatus, so that many components may be missing. Thus, for example, polymerizable substances are preferentially adsorbed.

With respect to 2:

(a) If we assume that all the sample has vaporized, that all the components have passed through the column packing without reaction or selective retardation and have finally reached the detector, where optimum conditions prevail, then we may say that the areas under the peaks represent a measure of the concentration of the appropriate components.

(b) The quantitative analysis of a mixture is often in spite of this easier to evaluate than the qualitative analysis.

(c) The quantitative result is affected not only by the mode of operation of the sample injector, the column, and the detector, but also strongly by the quality of the recorder and the method of evaluation.

There are thus many factors which work together to influence the qualitative and quantitative results. However, it is often quite unnecessary to carry out a complete qualitative and quantitative evaluation. With gas liquid chromatography we possess a whole series of new methods of presenting the results, which may be applied with complete safety.

We simply measure the relative height values, e.g. for the peaks, 1, 2, 3, ... n or even only for peaks 4 and 7 if these are characteristic for the process in question. We keep the operating conditions of our apparatus constant. Variations in the relative peak heights, i.e. h_7/h_4, are then true representations of variations in the process. For this peak 4 and peak 7 must obviously always appear after a time t_{r7} and t_{r4}, or in other words the ratio t_{r7}/t_{r4} must remain constant. This guarantees that variations in the process will always be recognized for the same components on the mixture studied. The chromatogram obtained has a high degree of reproducibility and without individual evaluation is a specific piece of information. It is only necessary to use it properly.

3.1. THE QUALITATIVE ANALYTICAL RESULT

FROM what has gone before it can be seen that practically any mixture can be separated, provided that the right column packing and the right operating conditions are used.

In the theoretical section it was shown that the retention time or volumes for whole homologous series could be precalculated. If the appropriate activity coefficients are known this is possible even for the most widely varied types of chemical compounds.

A commercial product, however, such as a ketone oil predistillate from the low-temperature carbonization of lignite, consists of at least 60 different substances. The number of possible isomers increases with increasing molecular size. It is necessary to have seen for oneself that e.g. a 0·5°C cut from a high efficiency distillation of C_8 olefins always contains 6–8 individual components, in order to appreciate this.

Normally, however, we are not interested in the structure of all these 6–8 components. Another matter of importance is the work that has been done prior to the analysis in order to improve the interpretation of the result obtained. The following example describes the process from the preparation of the sample to the qualitative and quantitative evaluation.

Task: determination of the solvent in a film emulsion.

Preparation of the Sample

The sample consists of volatile and non-volatile components. Although such mixtures can be directly injected into the gas chromatograph, it is better for accurate working to separate the substance into its volatile and non-volatile portions. If, for example, the mixture contains 50 per cent volatile material, 0·01 to 1 g of the mixture is spread over an inert surface (e.g. glass beads) in a U tube which is heated in an oil bath, a stream of ca. 3 l./hr N_2 is passed through, and the gas-vapour mixture is caught in a trap cooled with liquid nitrogen (see *Figure 96*).

The solvent mixture is now free from all solid matter; it has the same composition as the original sample, provided that the evaporation was quantitative. This is, however, only a matter of time. The nitrogen used as carrier gas must, of course, itself be purified by passing through a liquid nitrogen cooling trap, or else the solvent would, when frozen out, contain impurities from the nitrogen such as H_2O, O_2, CO_2; of these, liquid oxygen is particularly undesirable.

General Remarks

The aim of this preparatory work is to convert the substance to be analysed into a completely volatile mixture.

We can do this by means of the above gas distillation process, which is to be preferred to any other distillation process. It is so closely allied to gas liquid chromatography that the gas chromatographer will have all the required apparatus already. The gas distillation must be followed quantitatively, so that the original composition can be calculated later. On the other hand it is often a good thing to use suitable chemical or physical reactions to reduce the boiling points of the components, if these are too high, or to

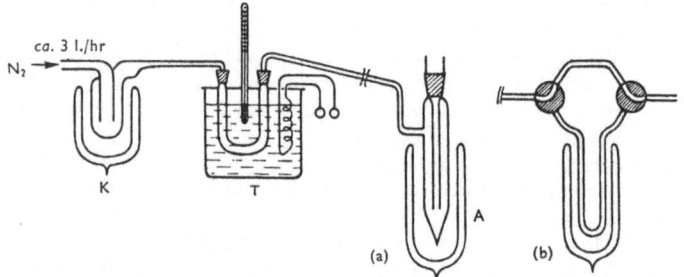

Figure 96. Apparatus for so-called gas distillation (principle of steam distillation with gas and addition of heat)

K = cooling trap for purification of the auxiliary gas; T = temperature control bath (the substance treated is in the U-tube in T); A = cooling trap; a = normal cooling trap for large amounts of substances; b = gas by-pass system for small amounts of substance.

increase the thermal stability of the products. Long chain fatty acids may be converted into their methyl esters with diazomethane, long chain alcohols are esterified with formic acid. Amino acids are converted at the carboxyl group or at the amino group into derivatives of lesser polarity, polymeric compounds are depolymerized, substances liable to hydrogenation are treated with hydrogen (carrier gas) over a little catalyst in the column, etc.

If, for example, the structure of hexantriol is to be determined, it is first treated with HI according to the methods of sugar chemistry to convert it into a hydrocarbon, and the latter is investigated. If in a mixture of oxidized compounds and hydrocarbons only the latter are of interest, the oxidized compound may be removed by a selective reaction.

Complex mixtures, such as are found in natural perfumes, may first be split up into their individual chemical groups. Such preliminary separations can, of course, be carried out with advantage by gas chromatographic methods, especially with multiple column instruments. The preliminary treatment of the sample must, in any case, only make use of definite reactions which involve no disproportionation. During such a treatment no substance should either appear or disappear in an uncontrolled manner.

Choice of the Most Suitable Column and Operating Conditions

From the data given under Columns in Volume III on the liquid phases suitable for a given separation, indications can be obtained of the way in which this problem may be tackled.

For our example, the analysis of the solvent for a film emulsion, the following points should be considered: the smell, refraction and general properties indicate that it is a complex solvent mixture which appears to consist of volatile esters and aromatic hydrocarbons. Alcohols may also be present. For a preliminary orientation we shall use two column packings, one strongly polar (polyethylene glycol) and one non-polar (paraffin oil). Of the two, polyethylene glycol shows itself as the more suitable, since the chromatogram obtained with it shows six symmetrical peaks, while that obtained from the paraffin oil column has only four peaks, one of which shows strong tailing (see *Figure 97* a, b).

General Remarks

Of the various columns which may be selected under the rules of polarity the one finally chosen is that which gives the best resolution for the greatest number of peaks. As a check, each individual peak is passed through a second column of different polarity in order to see whether any further breakdown is possible.

The temperature to be used depends partly on the required time of analysis and partly on the resolution of the substances appearing at the very start of the analysis.

The temperature may be raised until there is no further change in the number of peaks, i.e. there are no additional high-boiling substances appearing. With increasing temperature the retention times of the high-boiling substances are, in fact, decreased logarithmically. For such pre-

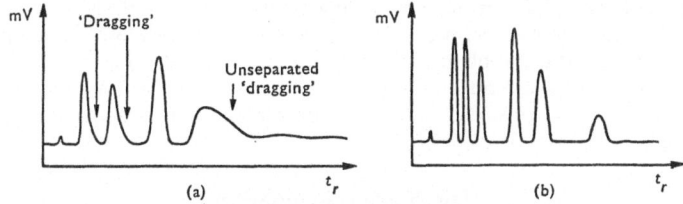

Figure 97. Choice of liquid phase
(a) liquid phase too weakly polar
(b) suitable liquid phase.

liminary investigations it is particularly useful to increase the column temperature stepwise, until the limiting operating temperature is reached (see *Figure 97* c).

The correct gas flow rate may be chosen when the optimum column performance is reached.

The smallest possible column length is that for which the resolution for the two substances most difficult to separate is just 50 per cent ($\vartheta = 0.5$).

As a rough working rule we can say:

The operating temperature should not be less than 60°C below the principal boiling point of the mixture; normally it should be 30°C below the final boiling point of the mixture. It should be noted that the resolution may alter with temperature and that the column performance generally falls with rising temperature.

Gas flow rate should be *at least* 1 ml carrier gas per minute for every square millimetre of column cross-section area, when optimum operating conditions are obtained.

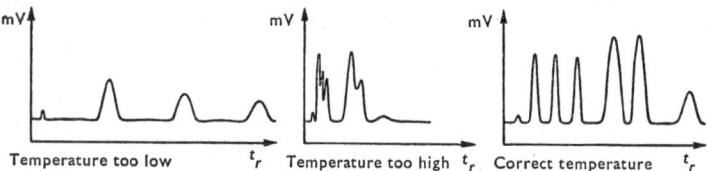

Figure 97c. Choice of column temperature.

For example, with a column of 4 mm internal diameter the cross-sectional area is 12·5 mm², and the gas flow rate must be at least 12·5 ml carrier gas/min. With a column of 6 mm internal diameter the cross-sectional area is 30 mm², the gas flow rate must be at least 30 ml carrier gas/min, measured under normal conditions at the column outlet.

The column length depends entirely on the separation to be carried out; short columns (1 to 2 m) should be used for high-boiling compounds and long columns (3 to 5 m) for low-boiling compounds. The decisive factor is that a resolution of 50 per cent should be obtained for the most difficult pair of substances. With a column of 2,500 plates it is possible to separate all the isomeric hexanes, almost all the isomeric heptanes, and at least ten of the isomeric octanes, provided that the most suitable liquid phase has been chosen.

Qualitative Evaluation

Returning to our example, we find that our unknown mixture gives a chromatogram as shown in *Figure 97* b.

The individual peaks may be identified as follows.

1. By their odour

As the components normally appear as individual substances, they may be identified by their odour at the column outlet (a glass tube is attached) provided that a thermal conductivity cell is used as detector and that each component leaves the column in a quantity of *ca.* 100 γ.

2. By selective freezing out and

    (a) chemical reactions
    (b) physical characteristics, n_D, E_p
    (c) spectroscopic investigations.

170

3. *By comparative analysis, i.e.*

by addition of possible substances and seeing whether the retention values are the same:

For our example, Method 1 gave the following:

Substances 1, 3 and 6 could not be identified. Substance 2 had the characteristic odour of acetate, substance 4 resembled benzene or toluene, substance 5 resembled butyl acetate.

These findings were confirmed by carrying out comparative analyses (Method 3).

(a) Addition of benzene.

Result: A seventh peak appears in the chromatogram between the original third and fourth peaks.

Conclusion: benzene is definitely absent.

(b) Addition of toluene.

Result: the fourth peak in the chromatogram is increased, but the odour at the end of the column is the same as before.

(c) The same thing was found for ethyl acetate and butyl acetate.

These two substances can be added simultaneously, since they are members of the same homologous series, and must therefore clearly be separated by a polar column.

The relative increase in the peaks for 2 and 5 on the addition of the two last-named substances is of course no proof that the peaks correspond to these substances; if, however, a new peak had appeared, this would have been definite proof that one of the two added substances had not been in the original mixture.

A similar odour *and* a similar retention time are, however, reasonably sufficient evidence that the added substance and the unknown substance are one and the same.

Substance 1 could be acetone. Acetone is added. The added component appears at the right place. However, there are doubts as to its odour.

Substance 1 is therefore frozen out. For this purpose a 200 mm long gas capillary tube is connected to the gas outlet behind the thermal conductivity cell. Shortly before the appearance of the substance the capillary is placed in liquid nitrogen.

After the recorder has shown that the peak for substance 1 has been passed, the base line has been reached and the line has started to rise for peak 2, the capillary is removed from the gas stream and a chemical micro-reaction is carried out with the frozen trace.

It is scarcely possible to identify the component which has been frozen out. But even a trace of acetone will, when added to a dilute solution of hydroxylamine hydrochloride faintly coloured with methyl orange, produce a red colour due to the oxime reaction.

If the reaction does not give a clear result a control test is carried out with added acetone. The amount of acetone added corresponds to the amount that might be contained in the mixture. The process is then carried out as above. The control test should now give a clearly positive reaction. If this is the case, then if acetone had been present in the original mixture the control reaction would also have had to give a positive result. The identification reaction should therefore be continued on a different type of column, or else processes other than merely the colour reaction should be used to determine the still unknown substance.

The reactions of substance 3 also did not give a clear result.

On freezing out, substance 6 is solid, and remains so until the capillary is defrosted, when it melts. A few grains of anhydrous copper sulphate (colourless) are introduced into the capillary. Under the microscope the grains can be seen to assume a pale blue colour. The addition of water to the original mixture increases the peak for substance 6 in the gas chromatogram. Substance 6 is thus probably water. If a little powdered calcium carbide is introduced into the top of the column and the mixture is injected, then if water is present a corresponding amount of acetylene should appear in the chromatogram.

The logarithms of the corrected retention times for substances 2 and 5 were plotted graphically; as 2 may be ethyl acetate and 5 may be butyl acetate the line shown in *Figure 98* was obtained, from which it can be clearly seen that substance 3 cannot be an *n*-acetate, but substance 1 lies on the straight line for the acetates.

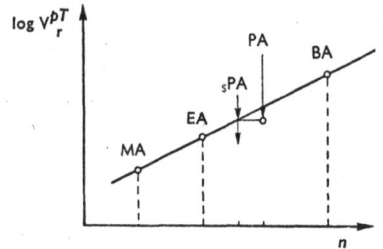

Figure 98. Study of relationships between members of a homologous series.

MA = methyl acetate
EA = ethyl acetate
sPA = sec-propyl acetate
PA = propyl acetate
BA = butyl acetate

The whole solvent sample is now treated with acetic anhydride, heated, washed with water, and neutralized with a saturated bicarbonate solution. The acetates are taken up in ether and the ethereal solution is chromatographed. The chromatogram shown in *Figure 99* is obtained.

Peak 1 has become slightly smaller. Peak 2 is still the same size as before. Peak 3, however, has now given rise to a new substance, which has a greater retention time than before esterification with acetic acid. A trace of peak 3 can still be seen in its original position. 3 is therefore an alcohol, which has now been converted into an ester. Nevertheless, its new retention time does not fit into the logarithmic relationship shown in *Figure 98*. It can therefore only be an iso-ester (e.g. isobutyl acetate) or the ester of a secondary alcohol (e.g. sec propyl acetate). Both sec propyl acetate and sec propanol are added to the esterified sample. The added substances show the same retention times.

Methyl acetate, which might form the first peak, is added and gives provisional confirmation of this fact. For full confirmation, it is necessary that the mixture containing added methyl acetate and the original mixture when passed through a column of different polarity should give completely

Figure 99. Control of the mixture for substances reacting with acetic anhydride (e.g. alcohols). After conversion and removal of surplus acetic anhydride the mixture is chromatographed; remnants of unconverted components and newly formed substances belong together.

sP = sec-propanol
sPA = sec-propyl acetate

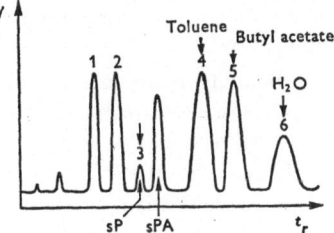

identical chromatograms. If this is found then the result may be taken to be confirmed.

Anticipating the results of the quantitative analysis, we build up an artificial mixture containing the same substances in the same proportions as are assumed to be in the original mixture. Infra-red spectrograms are then made of both the real and the artificial mixture. If the two are in perfect agreement, then the identification of the mixture may be regarded as correct.

General Remarks

The components may be identified by

(1) Application of direct measuring techniques to the previously separated, frozen out components, such as: infra-red spectrum, mass spectrum, n_D, E_p, elementary analysis, odour, chemical reactions etc.

(2) Comparison with test substances, which on two widely different columns must possess the same relative retention values (V_r, V_r^{pT}).

(3) Checking the functions
$\log t_r = a \cdot n + b$ or
$\log V_g = a' \cdot n + b'$ on one column, and
$V_g^{\text{rel I}} = c \cdot V_g^{\text{rel II}}$ on two columns with liquid phases I and II.
(a, a', b, b', c = constants, n = carbon number within a homologous series. See also Volume III.) 'Fractional odour control' is often sufficient when detectors are used that do not decompose the substance.
Rule: the higher the column performance, the more certain is the identification by the addition of test substances. The analysis of homologous series is particularly easy. For the exact determination of the molecular size it is only necessary to add one test substance. The identification is further simplified by

(4) Chemical conversion of the original substances to determine the chemical class.

Rule: the preliminary separation of the substance on chemically selective columns such as aliphatic/aromatic, alcohol/acid, monophenol/diphenol columns is to be recommended. Two column instruments may well be used for this purpose. The preliminary separation into groups of the same molecular size such as C_7 phenols/C_8 phenols, C_7 olefins/C_8 olefins simplifies the analysis of complex mixtures of isomers. The identification is further simplified by

(5) Simplifying the complex mixture by

    (a) extractive treatment with conc. H_2SO_4, KOH, silica gel, urea adduction, etc.

    (b) hydrogenation

    (c) saponification, esterification

    (d) etherification, which should reduce the number of types of chemical compound in the mixture, etc.

(6) Use of tabulated values of the specific retention volume V_g, the partition coefficient K, the relative retention value or the retention index I (Volume III).

Preliminary chemical treatments can also be carried out during the direct analytical process. For example, a selective reaction column may be connected directly behind the sample injector. If such a column is filled with silica gel saturated with oleum, it will retain all the unsaturated and most of the oxidized compounds, together with all basic substances and many aromatics.

The efficiency of a reaction column may be tested by carrying out one analysis with and one analysis without it.

The result of such a test also helps to identify the substance being analysed. This process is called difference analysis.

The small quantities of substance normally isolated are sufficient for the determination of physical characteristics. For example, only γ quantities are required for a mass spectrum (see Bradford[4]). Recently infra-red spectrometer cells have been produced which require only 0·002 ml liquid.

James and Phillips[5] describe a simple micro-Schleiermacher apparatus with which they were able to determine boiling points to 0·1°C on 0·02 ml liquid (*Ber. dtsch. chem. Ges.* 1891, **24**, 944).

Littlewood[6] determined molecular weights of 10 to 100 γ samples to an accuracy of $\pm$ 5 per cent in an effusiometer. The molecular weight may be determined more elegantly and exactly by gas chromatography. For this the unknown substance is mixed with a known quantity of a component which may be separated from it. Now two analyses are carried out one after the other with two gases of very different molar weights (e.g. H_2 and CO_2). The molar weight of the unknown substance may be determined from the areas of the peaks obtained in each case (see Liberti[7]). The gas density balance is especially suitable as detector for such determinations.

More recently, Walsh and Merritt[8] have developed a method worthy of wider use for ensuring qualitative analytical results from mixtures of complex chemical composition. The method is essentially a combined evaluation of retention values and specific chemical colour reactions. The gas stream from the column after passing through the (gastight) detector is led simultaneously into solutions of different reagents. It passes through a three-way tap into a plastic cap in which about five hypodermic syringes are sticking in such a way that the gas stream can be passed into five small test-tubes simultaneously, each of which is filled with a specific reagent. Walsh and Merritt suggest the following reagents:

For alcohols (primary and secondary; tested for C_1–C_4)
0·5 ml 1·5N HNO_3+1 drop 1 per cent $K_2Cr_2O_7$
Colour change from pale yellow to blue-grey. Minimum quantity of component detected: 20 γ.

For aldehydes (tested for C_1–C_6)
10 drops 2,4 dinitrophenylhydrazine
Yellow-orange ppt. Minimum quantity of component detected: 20 γ.
10 drops freshly prepared Schiffs reagent.
Colour change from colourless to purple. Minimum quantity: 50 γ.

For ketones (tested for C_1–C_6 methyl ketones)
10 drops, 2,4 dinitrophenylhydrazine.
Yellow ppt. Minimum quantity: 20 γ.

For esters (C_1–C_5 acetates tested)
10 drops of 1N NH_4OH . HCl in methanol+3–4 drops of 2N alcoholic potash (until the solution goes blue). After passage of the gas add 5–6 drops of 2N HCl with a pipette until the solution becomes clear and colourless. Then add 1–2 drops of a 10 per cent aqueous solution of $FeCl_3$. Colour change from colourless to red. Minimum quantity: 40 γ.

For alkyl halides (tested for C_1–C_4)
10 drops of 2 per cent alcoholic $AgNO_3$ solution.
White ppt. Minimum quantity: 20 γ.

For amines (tested for C_1–C_6)
5 drops of pyridine +1 drop of 5 per cent NaOH.
After passage of the gas add 1–2 drops of benzenesulphonyl chloride. Colour change from colourless to yellow for primary and secondary amines. Colour change from colourless to pink to deep purple for tertiary amines. Minimum quantity in all cases: 100 γ.
10 drops of water +2 drops of acetone +1 drop of 1 per cent sodium nitroprusside solution. If necessary add 1–2 drops of acetaldehyde.
Red coloration with primary amines.
Blue coloration (after addition of acetaldehyde) with secondary amines.
10 drops 1N NH_4OH . HCl in propylene glycol +2 drops 1N KOH in propylene glycol. After passage of the gas the mixture is rapidly heated to boiling and cooled. The solution must be clear and colourless. Add 1–2 drops of 10 per cent $FeCl_3$ solution. Reddish-blue coloration. Minimum quantity: 40 γ.

11

For mercaptans (tested for C_1–C_7)
10 drops of 2 per cent alcoholic $AgNO_3$ solution.
White ppt. (black in presence of H_2S).
Add 10 drops of saturated alcoholic Pb acetate solution.
Yellow ppt. Minimum quantity: 100 γ.
10 drops of 1 per cent isatin in conc. H_2SO_4.
Green coloration (also for alkyl disulphides). Minimum quantity: 100 γ.

For alkyl sulphides (C_1–C_8)
10 drops of 95 per cent ethyl alcohol +2 drops of 5 per cent KCN in 1 per cent NaOH solution. 2–3 minutes after passage of the sample add 5 drops 1 per cent sodium nitroprusside solution.
Colour change to red. Minimum quantity: 50 γ.

For aromatics and unsaturated aliphatics
10 drops conc H_2SO_4+1 drop 37 per cent HCHO solution.
Colour change to reddish blue. Minimum quantity: 20–40 γ.

These colour reactions may be found in *Semimicro qualitative organic analysis* by Cheronis and Entrikin, 2nd ed., Interscience, New York, 1957, and in *The systematic identification of organic compounds* by Shriner, Fuson and Curtin, 4th ed., Wiley, New York, 1946. This principle may be applied to any desired type of chemical substance, even though the molecular sizes of the substances which can be definitely identified may be limited. The specific chemical colour reactions may even be used when there are several compounds in one peak. Once the chemical class has been determined the retention values will give the molecular size.

Qualitative Testing of Columns

Before a column is used for the analysis of completely unknown mixtures a great deal should be known about its selectivity. This may be accomplished as follows.

Suppose that the column is suitable for the separation of alcohols, esters and aromatics.

The optimum gas flow rate is adjusted or determined by continually injecting the same amount of a simple mixture, such as butyl acetate/toluene, and varying the gas flow rate at constant temperature from 0·5 to 5 l./hr. The separation values obtained at each gas flow rate are plotted as $t_r/b_{\frac{1}{2}}$ or n' against the gas flow rate. The function of $t_r/b_{\frac{1}{2}}$ against F shown in *Figure 100* is obtained. The optimum value in l./hr is now obtained by graphical interpolation.

Now the column is tested with test substance a and the chromatogram is recorded. An equal weight of test substance b is now mixed with it (e.g. 0·5 ml in each case); naturally the total amount used does not exceed the load capacity of the column.

The stepwise increase in the test mixture leads to a series of chromatograms which can be interpreted completely unambiguously, since even superimpositions can be recognized by the doubling of the area of the peak concerned (see *Figure 101*).

Rule: the testing of columns for qualitative analysis is carried out by the use of mixtures which are built up in steps, at each step an equal quantity of a new substance being added to the mixture. Between each addition a

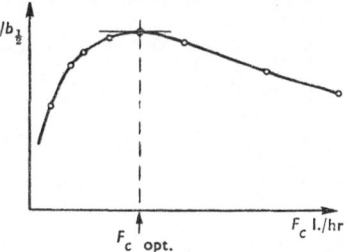

Figure 100. Determination of the optimum gas flow rate F_c.

chromatogram is obtained. Superimpositions may be recognized by the fact that the area under the peak concerned is two or more times as great as would be expected. Column testing is only valid over a narrow temperature range.

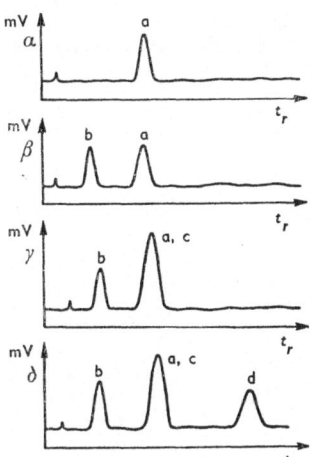

Figure 101. Testing the chromatograph for specific separating properties (selectivity and column performance). Each component in the test mixture, which is built up stepwise, is represented by the same concentration.

The identification simply by means of comparative values from the literature is critical. The most exact information on the way in which values from tables $(r, t_r^{rel}, I, V_g, K)$ may be used with certainty for identification purposes will be found in Volume III. Data on the way that hydrocarbon mixtures may be identified by stepwise synthesis of isomeric hydrocarbons from the lower hydrocarbons by the diazomethane method will be found in Volume II.

References

1. JAMES, A. T. and MARTIN, A. J. P., *Biochem. J.*, 1952, **50,** 679.
2. AMBROSE, D., KEULEMANS, A. I. M. and PURNELL, J. H., *Analyt. Chem.*, 1958, **30,** 1582.
3. LITTLEWOOD, A. B., PHILLIPS, C. S. G. and PRICE, D. T., *J. Chem. Soc.*, 1955, 1480.
4. BRADFORD, B. W., HARVEY, D. and CHALKLEY, D. E., *J. Inst. Petrol.*, 1955, **41,** 80.
5. JAMES, D. H. and PHILLIPS, C. S. G., *J. Chem. Soc.*, 1953, 1600.
6. LITTLEWOOD, A. B., *Analyst*, 1956, **81,** 53.
7. LIBERTI, A., CONTI, L. and CRESCENZI, V., *Nature, Lond.*, 1956, **178,** 1067.
8. WALSH, J. T. and MERRITT, C., *Analyt. Chem.*, 1960, **32,** 1378.

Further Literature on Qualitative Gas Chromatography

RUSSELL, D. S., *Canad. J. Chem.*, 1958, **36,** 1745, Gas chromatography and infra-red spectrography; and in the corresponding sections of *Gas Chromatography Abstracts*, ed. C. E. H. Knapman, Butterworths, London (1958, 59, 60 and following issues).

ANDERSON, D. M. W., *Analyst*, 1959, **84,** 50, 5 micromoles of vapour of liquid are sufficient for an infra-red determination.

GOHLKE, R. S., *Analyt. Chem.*, 1959, **31,** 535.

WEYERMULLER, G., *Chem. Processing*, 1959, **22,** 107, Gas chromatography and mass spectrometry.

3.2. THE QUANTITATIVE RESULT

BEFORE we deal with the limitations which must be borne in mind, let us take our example a stage further.

The first step is to determine the areas under each of the peaks 1 to 6. This may be done to a first approximation by multiplying the peak height by the corresponding retention time and adding the products:

$$h_1 . t_1 + h_2 . t_2 + h_3 . t_3 + \ldots + h_6 . t_6 = \sum_1^6 A_1 \ldots {}_6$$

Further we can say with certain reservations that the relationship of the individual areas to the sum of the areas is the same as that of the weight percentages of the corresponding substances to 100.

$$\frac{h_1 . t_1}{\sum A_1 \ldots {}_6} = \frac{\% \, 1}{100} \; ; \; \% \text{ substance } 1 = 100 . \frac{h_1 . t_1}{\sum A_1 \ldots {}_6}$$

In the same way the percentage values for 2 . . . 6 may be obtained and it is now possible to synthesize a mixture corresponding to these values. The synthetic mixture is, in its turn, analysed, and the chromatogram obtained is compared with the one for the natural mixture. There are, however, a number of difficulties attached to this somewhat intricate method, even though it may at this point be stated that the complete agreement of the two chromatograms constitutes proof of the quantitative identity of the two mixtures.

We must therefore leave our example at this point and deal with the problem in a more general manner.

Requirements for the Quantitative Evaluation of Gas Chromatographic Results

1. The detector used must have a linear signal. This may be tested by injecting increasing amounts of a uniform mixture into the chromatograph.

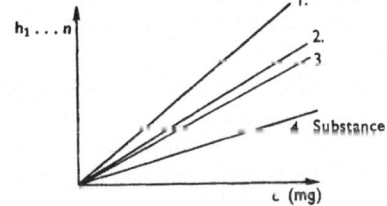

Figure 102. Calibration curves.

The individual peak heights are measured and plotted against the corresponding quantities of substance. The operating conditions of the instrument are kept completely constant during this test. The relationship $h_i/c_i =$ const. shown in *Figure 102* must be obtained. The smaller the quantities

179

of substance that are used, the more completely is the requirement met. A considerably wider range of linearity is obtained if the peak area is plotted against the amount of substance, since then the influence of the somewhat limited capacity of the column is not superimposed upon the properties of the detector itself. In this way the linearity of the detector may be more surely recognized.

2. The recorder must have a linear signal, but at the same time it must also have an absolutely constant chart speed. This may be checked by connecting a source of linearly varying potential to the recorder input and recording a time/voltage curve. This must be a straight line.

If these two requirements are fulfilled, then the area under a peak, in other words the integral

$$\int_{t_0}^{t_e} A \cdot dt$$

corresponds to the quantity of substance.

This is not to say that equal areas for different substances correspond to equal amounts.

It also does not answer the question of whether the areas represent mole or weight percentages.

Measurement of Peak Area

1. *By Electronic Integration*

Following a suggestion from Geppert, the author, in collaboration with Kuhl and with the assistance of Renker, has developed an electronic integrator which breaks the peak areas down into impulses which can be added up. For this *ca.* 0.1 mm^2 surface area corresponds to one electronic impulse. With a counting rate of 500 impulses per sec ($= 500$ c/s) even the largest peaks on a chromatogram can be evaluated with a precision that is greater than the precision obtained from the electronic measurement of the detector signal. (There are many types of electronic integrator.)

The peak area can thus be read off directly at any time during continuous analysis, or the values can be printed if a print-out device is connected to the counting mechanism. The values can also be recorded directly as voltages (Anna[1]).

2. *By Electro-Mechanical or Mechanical Integration*

The detector signal is amplified and the current produced is fed to a current counter or integration motor. The revolutions of the counter are detected optically; the number of revolutions or a multiple thereof is recorded by an electromechanical counter and printed.

3. By measuring the height h and the corresponding width at half height $b_{\frac{1}{2}}$ and multiplying the two values together a good approximation $h \cdot b_{\frac{1}{2}}$ is obtained. According to Jaulmes and Mestres, *Compt. Rend.*, 1959, **248**, 2752, a more accurate value is obtained by multiplying the peak height

with the width at 45·5 per cent height. However, as this method is really only intended to give approximate values, the extra complication seems hardly worth the trouble. It is also possible to use a planimeter, but this gives neither rapid nor very accurate results.

4. By multiplying the peak height h by the retention time t_{dr} another approximate value is obtained, but this is not as good as the one obtained by method 3.

The error in peak area measurements by $h \cdot t_{dr}$, as compared with that obtained from $h \cdot b_{\frac{1}{2}}$, is anything up to $\pm$ 10 per cent (see Cheshire and Scott[2]).

The greater the number of peaks, the better is the result obtained by this method. A gasoline analysis with over 140 peaks gives a good result when evaluated by this method, but a mixture of only three components cannot be evaluated at all, since in such a case the error may exceed 30 per cent.

5. Finally the peak area can be obtained by cutting out the peaks and weighing them.

In many cases this somewhat demanding process cannot be avoided, especially where a trace of substance on the rear flank of a large peak is to be measured. The determination of the correct area is simplified if a column of particularly good separating power is used.

A poor base line and insufficient separation affect the quantitative result obtained by the various methods as follows:

Adverse Effect of Poor Base Line and Poor Separation	*Method of Measurement*
Very strong	1
Very strong	2
Moderate	3
Weak	4
Hardly at all	5

1. Evaluation from Peak Area Measurements. The Direct Method

The detector has a linear signal; it will give the same deflection for similar concentrations of different substances. This assumption is approximately correct when chemically similar substances are to be detected by the thermal conductivity method using hydrogen or helium as carrier gas.

In this very simple case, we may say:

$$\text{Substance 1 (wt. per cent)} = \frac{A_1 \cdot 100}{\sum\limits_{1}^{n} A} \qquad ....(1)$$

181

A_1 = Peak area of substance 1

$\sum\limits_{1}^{n} A$ = sum of the peak areas for all substances

Note: in the analysis of homologous series the thermal conductivity method is particularly favourable for quantitative evaluation.

The validity of the basic assumption may be tested where necessary by analysing a mixture of known composition and seeing if any deviations occur. Suppose the test mixture has the composition b wt. per cent of substance i; c wt. per cent of substance k.

The chromatographic surface areas

$$A_i/A_k \text{ should be as } b/c$$

i.e. $A_k \cdot b = A_i \cdot c$.

If the deviation is greater than 1 per cent relative, then the assumption is not valid, and the simple formula 1 cannot be used.

2. Evaluation by Peak Area Measurements using Specific Calibration Factors

The detector has a linear signal. Equal concentrations of different substances in the carrier gas do not, of course, produce equal deflections; however, the value for the deflection varies linearly with concentration, and also depends on a specific property of the substance concerned.

By way of example, let us consider the micro flame ionization detector, the micro flame detector, the β-ray ionization detector for measurements within homologous series, and the thermal conductivity cell for measurements outside homologous series. The heat of combustion of the substance under investigation influences the deflection of the micro flame detector. The carbon content of the substance influences the deflection of the micro flame ionization detector. The ionization cross-section (i.e. for homologous series the empirical formula) has a partial influence on the deflection of the β-ray ionization detector.

In all these cases the surface area A_i of the individual substance must be corrected with a factor f_i which is specific for i, so that the correct quantitative result may be obtained:

$$\text{Substance 1 (wt. per cent)} = \frac{f_1 \cdot A_1 \cdot 100}{\sum\limits_{1}^{n} f_{1,n} \cdot A_1 \cdots n} \qquad ....(2a)$$

$f_1, f_2 \ldots$ are correction factors, for which on each occasion specific values must be inserted corresponding to the type of detector used.

For the micro flame detector we can say approximately:

f_1 = 1/heat of combustion of substance 1
f_2 = 1/heat of combustion of substance 2, etc.

For the micro flame ionization detector:

$f_1 = 1/$per cent carbon content of substance 1
$f_2 = 1/$per cent carbon content of substance 2 etc., or molar weight of substance 1/carbon number . 12

For the thermal conductivity cell:

$f_1 = 1/$thermal conductivity of substance 1
$f_2 = 1/$thermal conductivity of substance 2 etc.

Naturally the correction factors do not always have to be determined specially, but can be calculated from values (such as the heat of combustion) which are already known.

In any case, certain simplifications may also be made:

In the analysis of homologous series of hydrocarbons with a micro flame detector, the correction with the reciprocal heat of combustion may be omitted if the results are given not as weight but as volume percentages. The heats of combustion produced on the combustion of equal volumes of hydrocarbons within a homologous series are independent of the molar weights of the individual substances.

In this case the specific correction factors are always the same and thus do not enter into the calculation. It should be remembered that in this case the formula is:

$$\text{Substance 1 (vol. per cent)} = \frac{A_1 \cdot 100}{\sum\limits_{1}^{n} A} \qquad \ldots(2b)$$

The values are thus obtained as volume percentage values.

Mole Percentages or Weight Percentages?

This question has been analysed by, among others, Heft[3]. The data obtained relate only to the case where a thermal conductivity cell is used as detector. When hydrogen or helium is used as carrier gas we may say that the uncorrected peak areas within a homologous series are related to the weight percentages. In the analysis of mixtures other than of homologous compounds, the areas are not related to the weight percentages, but they are even less related to the mole percentages.

A correction using thermal conductivities, as recommended above, reduces the error in the case where the combined thermal conductivities between the substance and the carrier gas vary linearly with the concentration.

If as carrier gas argon, nitrogen or carbon dioxide is used, the areas are only approximately related to the weight percentages in the case where the members of a homologous series have the same specific heats in the vapour phase.

Mole weight percentages are thus always, with a few exceptions, liable to error.

Similar results were obtained by Browning and Watts[4], Dimbat, Porter and Stross[5] and Fredericks and Brooks[6].

The widespread but erroneous assumption that, particularly in the case where a thermal conductivity cell is used as detector, the mole percentage concentrations can be calculated directly from the peak areas, has also been refuted by the work of Hinkle and Johnsen[7], who found that values for the vapour pressures of the substances must be used as correction factors if the results are to be obtained as mole percentages:

$$\text{Substance 1 (mole per cent)} = \frac{A_1 . P_1^{-\frac{2}{3}} .100}{A_1 . P_1^{-\frac{2}{3}} + A_2 . P_2^{-\frac{2}{3}} + \ldots + A_n . P_n^{-\frac{2}{3}}} \quad \ldots.(2c)$$

for which the correction factors are

$$f_1 = P_1^{-\frac{2}{3}}$$
$$f_2 = P_2^{-\frac{2}{3}} \qquad \text{etc.}$$

and are measured at the same temperature.

This empirical correction has been derived from the behaviour of the peak area when the pressure conditions fluctuate during the measurement.

3. Evaluation of the Analysis by a Calibration Method. External Calibration of the Peak Heights

Especially in cases where not all the peak areas can be recorded on the chromatogram, where the evaluation must be carried out very quickly, or where the mixtures are so complex that a simple correction such as that given in the previous section would not be possible, the chromatogram is evaluated by the so-called external calibration method.

The prerequisite for this method is that the apparatus must always be run under the same conditions. The temperature of the column, the sample injector and the detector must either be kept absolutely constant between calibration and measurement, or else be capable of being adjusted to give the same value. The gas flow rate, the pressure drop along the column, the column performance and the detector conditions must also be kept absolutely constant. If such conditions are maintained then the quantitative analysis can be evaluated by the peak height method.

The first step is to prepare calibration diagrams. Varying amounts of the substance are injected. The peak heights read off from each of the chromatograms are measured and plotted against the amounts of substance injected. If the mixture contains n components, then n calibration curves are prepared. The calibration curves are normally linear, as long as only very small amounts of substance are used which do not exceed the load capacity of the column (about 3 mg per component for 6 mm columns with a normal quantity of liquid phase).

If the quantity of substance cannot be varied, then the same quantities of calibration mixtures of different composition are injected.

Figure 102 shows what such calibration diagrams look like. If now the above conditions are fulfilled it is possible, by measuring the peak height h_i and comparing it with the calibration diagram i, to establish how many mg or ml of the mixture were injected, in order to be able to calculate the wt. per cent or vol. per cent in the mixture.

Where the calibration curves are straight lines, whose gradients depend on the retention times of the corresponding substances in the column, we may say

$$h_i = k_i \cdot w_i$$

h_i = peak height; w_i = amount of i used for calibration in mg or ml, or

$$w_i = h_i/k_i \text{ (mg or ml)}$$

Let the amount of substance injected be e (ml or mg). Then the content of i in the mixture is:

$$\text{Substance } i \text{ (wt. per cent or vol. per cent)} = \frac{h_i \cdot 100}{k_i \cdot e} \quad \text{....(3)}$$

k_i is called the external calibration factor.

If the calibration curves are not straight lines then the detector is not giving a linear signal, the column is overloaded, or the liquid phase has non-ideal behaviour. In such cases the formula becomes

$$\text{Substance } i \text{ (wt. per cent)} = \frac{100}{e} \cdot \text{Calibration value corresponding to } h_i$$

e = amount weighed out in mg.

The peak heights in the chromatogram must have no influence on one another. The column resolution must be at least $\theta \leqslant 50$ per cent.

4. External Calibration of the Peak Areas

The requirements for this process are considerably less exacting than those for the peak height method. The temperature, type of column and the column performance may all vary, which is especially important in the case of multi-column instruments. Only the detector and its operating conditions together with the gas flow rate must be kept constant, then we can say that the peak area of a substance is equivalent to its quantity.

The calibration diagrams are prepared either by injecting varying amounts of substance or by injecting constant amounts of mixtures of varying composition, and plotting the areas obtained in the chromatogram against the corresponding amounts of substance.

The calibration curves are straight lines; they may have the same gradient. We may say

$$\text{Substance } i \text{ (wt. per cent)} = \frac{A_i \cdot 100}{k_i \cdot e}$$

if e is given in mg and the w_i values are also given in mg. If e is given in volume units and w_i is used in the same volume units for calibration, then the vol. per cent may be obtained from i.

When a thermal conductivity cell is used as detector then differences in the k values are due to strongly differing thermal conductivity values, if H_2 or He are used as carrier gas. They are also traceable to differences in the specific heats when CO_2, A or N_2 are used as carrier gas. It is of course necessary to use the same carrier gas for the calibration as for the actual analysis.

When a detector other than a thermal conductivity cell is used, the calibration curves are still straight lines, but their gradients, and thus their k values, may vary.

The *advantage* of the external calibration method of evaluation is that only those peaks corresponding to substances of interest need be evaluated in order to obtain the information desired. This means that it is neither necessary for the entire chromatogram to be recorded, nor need the entire mixture be volatile. Among other things, this is of importance for trace analyses where the main component cannot be detected with accuracy, or when analyses are to be performed on substances such as lacquers, glue, gas in oils, gas in water, etc.

The *disadvantage* is that the substances to be detected must already be available as individual substances, so that the calibration diagrams may be made. However, this problem may be solved with the help of preparative gas chromatography.

5. Internal Calibration with addition of Extraneous Substance

In the so-called internal calibration, a known amount of a substance not present in the original mixture is added to the mixture. The peak area of the added substance is determined from the chromatogram and is compared with the area of a substance of interest. If equal amounts of the added substance and the substance of interest i produce equal peak areas, then the calculation is simple:

$$\text{Substance } i \text{ (wt. per cent)} = A_i \frac{100g/(z + g)}{A_F} \qquad ....(5)$$

A_F peak area of the added substance
A_i peak area of component i
z weight of impurity in mg added to
g mg of original mixture

6. Internal Calibration with Component Already Present

If, however, the peak areas A_i and A_F produced by equal amounts of i and F are not equal, then the internal calibration must be carried out with the substance already present in the mixture. In this case equal amounts do produce exactly equal areas.

It is now no longer possible to determine the area for the added substance and relate it to the amount added; rather one has to determine the increase in A_i caused by the addition of i. This somewhat complicated procedure provides the surest way of quantitative measurement, provided that the detector has a linear signal and that its operation can be kept absolutely constant.

The procedure is as follows.

(1) A chromatogram of the mixture is prepared.

(2) With the conditions kept constant, a chromatogram of the substance to be added is prepared, and the purity of the additive is obtained as a percentage.

(3) The additive is added to the original mixture, the amount added in wt. per cent being exactly known.

(4) A chromatogram of this mixture is prepared.

(5) The result is calculated by formula 6.

Note: as the amounts of substance used for the chromatogram do not necessarily have to be the same, the area of a peak which is not affected by the added substance or any impurities it may have (see above under 2) is used to convert the figures to equal quantities injected.

$$\text{Substance } i \text{ (wt. per cent)} = \frac{100 . A_i . z_i}{\left(\frac{A_x^{I}}{A_x^{II}} . A_{zi} - A_i\right)\left(100 - z_i\right)} \quad \text{....(6)}$$

where

A_{zi} = area of peak i after addition of z per cent of i in analysis II

A_i = area of peak i without added substance in analysis I

z_i = added substance i as wt. per cent of newly formed mixture which is put equal to 100

A_x = area of any unaffected peak x in the first analysis

A_x^{II} = area of the same peak x in the second analysis.

7. Evaluation by the Height and Area Methods for Constant Sample Quantity

If the volume of sample injected is always the same, then the heights or the areas are a measure for the volume concentration of the individual substances, provided that the operating conditions of the apparatus, and in particular those of the detector, are kept constant.

Particularly in the case where the products to be analysed are always the same, evaluation with 'linear percentages' is possible. These are lines marked with specific scales for each component, which are held against the peak heights for the components. The concentration can be read off directly. Such methods are only of value when the product under investigation is always the same, and thus they are used for the gas chromatographic control of commercial production units (Fischer[0]).

Naturally, this method also requires quantitative calibrations.

Evaluation Methods using Integral Detectors

The quantitative evaluation of chromatograms obtained from integral detectors is so simple that it need not be gone into in any great detail here. It is a well-known fact that the step height in an integral chromatogram is a measure for the amount of substance corresponding to the step. Two examples should suffice.

1. Suppose that the integral detector is an automatic burette and the analysis is being carried out on monobasic fatty acids. The step heights of the chromatogram are directly proportional to the volumes of titration fluid used. The step height may be calculated from the volume of titration fluid used (in ml) by means of the equation:

$$h = k \cdot a$$

h = step height in mm

a = titrant in ml

k = linearity or amplification factor of the detector, which is independent of substance.

The amount of a monobasic acid i may be obtained from the equation:

$$g_i = M_i \cdot \frac{h_i}{k} \cdot f_{\text{titrant}} \cdot n \ (\text{mg})$$

g_i = weight of acid i in mg

M_i = molecular weight of acid i

h_i = height of step corresponding to i

f_{titrant} = titrant factor

n = normality of titrant (e.g. 0.1 N)

k = linearity factor of the detector.

$$k = \frac{h_i}{a_i} = \frac{\text{step height for substance } i}{\text{ml titrant used for } i}$$

The weight percentage of i may be calculated from the well-known equation:

$$\text{Substance } i \ (\text{wt. per cent}) = \frac{100 \cdot g_i}{e} \qquad \ldots(7)$$

e = amount of substance in mg weighed out per analysis

2. Consider the case of a Janak detector modified according to the proposals of Leibnitz and Hrapia[9]. The step height corresponds to the partial volume of the gas component i.

$$\text{Substance } i \ (\text{vol. per cent}) = \frac{\frac{h_i}{k} \cdot 100}{e} \qquad \ldots(8)$$

$h_i = k \cdot v_i$

h_i = step height in mm

v_i = partial volume in ml

e = volume of gas mixture injected in ml

Conclusions

A comparison of the methods used for the evaluation of gas chromatographic quantitative analyses with those used e.g. for the evaluation of spectrographic analysis shows that the former are far more simple. As, however, the different evaluation processes used in gas liquid and gas solid chromatography have different requirements, a general survey is given in Tables 1 and 2.

Table 1

Influence of variations in operating conditions on the quantitative result (error)

Method	Δ Temperature	Δ Gas Flow Rate	Variations	Work output needed
1	Very slight	Weak	Slight	Medium
2	Very slight	Weak	Slight	Great
3	Very slight	Weak	Slight	Medium
4	Slight, but temperature at point of injection must be kept absolutely constant	Strong	Strong	Medium to Slight
6	Very slight	Weak	Slight	Medium
7	Very slight	Very slight	Slight	Medium

Table 2

Constancy of operating conditions required for reproducible qualitative and quantitative analysis with the thermal conductivity cell.

Δ gas flow rate:	$< \pm$ 0·5 per cent
Δ column temperature:	$\pm$ 0·1°C for qualitative analysis
	$\pm$ 10°C for quantitative analysis
Δ column pressure:	$\pm$ 0·5 per cent
Δ column length:	$\pm$ 10 per cent

The operating conditions of the thermal conductivity cell must be kept constant as follows:

For qualitative analysis: any normal variations

For quantitative analysis:

Cell pressure: $\pm$ 1 mm at 760 mm during measurement and calibration

Heating current: $\pm$ 0·1 per cent

Cell temperature: $\pm$ 1°C, if the heating current is kept constant

Maximum accuracy of the analytical result:

$\pm$ 0·2 per cent of 100 per cent component, if the result is registered by a recorder which does not produce any additional errors.

References

1. ANNA, O., *Elektron. Rdsch.*, 1958, **4**, 127.
2. CHESHIRE, J. D. and SCOTT, R. P. W., *J. Inst. Petrol.*, 1958, **44**, 74.
3. LEIBNITZ, E., KAISER, R. and HEFT, C., *Gas-Chromatographie 1958*, ed. H. P. Angele, Akademie-Verlag, Berlin, 1959, 59.
4. BROWNING, L. C. and WATTS, J. O., *Analyt. Chem.*, 1957, **29**, 24.
5. DIMBAT, M., PORTER, P. E. and STROSS, F. H., *Analyt. Chem.*, 1956, **28**, 290.
6. FREDERICKS, E. M. and BROOKS, F. R., *Analyt. Chem.*, 1956, **28**, 297.
7. HINKLE, E. A. and JOHNSEN, S. E. J., *Gas Chromatography*, eds. Coates, Noebels and Fagerson, Academic Press Inc., New York, 1958, p. 25.
8. FISCHER, J., *Gas-Chromatographie 1958*, ed. H. P. Angele, Akademie-Verlag, Berlin, 1959, 22.
9. LEIBNITZ, E. and HRAPIA, H., *Gas-Chromatographie 1959*, ed. H. P. Angele, Akademie-Verlag, Berlin, 1959, 144.

3.3. APPLICATION OF GAS CHROMATOGRAPHY TO THE DETERMINATION OF THERMODYNAMIC VALUES

As HAS already been mentioned, gas chromatography can often be used as a simple and exact method for the determination of thermodynamic values. Thus Kozub[1] was able, by exact measurements on the system methanol/methyl formate in an equilibrium apparatus, to obtain values of astoundingly high precision for the vapour/liquid phase equilibrium.

Anderson[2, 3], Hoare and Purnell[4], Keulemans[5], Kwantes and Rijnders[6], Littlewood, Phillips and Price[7], Porter, Deal and Stross[8] and White and Cowan[9] have shown the value of gas liquid chromatography for the determination of thermodynamic functions.

Activity coefficients and values for the free energy of mixing can be determined from the retention values at a given temperature.

Heats of mixing can be calculated from the temperature dependence of the retention values.

Values for the entropy of mixing may be calculated from the heat of mixing and the free enthalpy of mixing.

If the solid support is not a substance like kieselguhr or clay, which are too active and tend to take part in the retention, but instead a substance such as steel in the form of very fine turnings, exact results may be obtained.

Kwantes and Rijnders[6] extended their measurements to include relatively volatile liquid phases. They saturated the carrier gas stream with the vapour of the liquid phase and were thus able to determine the activity coefficient of a substance in its closest homologue (e.g. of heptane in octane).

Determination of the Activity Coefficient

γ° as γ°_{PT} at a vapour pressure of solution p° and at a temperature T from the following calculation:

$$\gamma^{\circ}_{PT} = \frac{N_L \cdot R \cdot T}{K \cdot p^{\circ}}$$

N_L = moles of liquid phase per ml at temperature T

p° = vapour pressure of liquid phase under normal conditions

K = partition coefficient, obtained from

$$K = \frac{V_g \cdot \rho L \cdot T_s}{273}$$

For exact measurements of activity coefficients the operating conditions of the apparatus must be ideal as follows:

very small amount of substance
high precision measurement of retention time

12   191

consideration of the dead volume of the column, the detector and the connecting parts

exact measurements of all values needed for the determination of the retention volume and corrections

long column (at least 6 m)

exact measurement of the density of the liquid phase at all temperatures and careful correction for weight loss due to evaporation in carrier gas.

Values for activity coefficients determined in this way are given by Keulemans[5], by Kwantes and Rijnders[6] and by Wolf and Beyer[10].

Determination of the Heat of Mixing, the Free Energy of Mixing, and the Entropy of Mixing

The partial molar heat of mixing ΔW may be obtained by plotting the logarithm of the partition coefficient K against the inverse of the absolute temperature.

$$\frac{d \ln K}{d\left(\frac{1}{T}\right)} = - \frac{\Delta W}{R}$$

ΔW = partial molar heat of mixing
R = gas constant

The entropy of mixing, the free enthalpy of mixing and the heat of mixing are, by definition, related in the following way:

$$\Delta S = \frac{\Delta W - \Delta G}{T}$$

The value for the free enthalpy of mixing may be obtained from the equation

$$\Delta G = - R \cdot T \cdot \ln \frac{273 \cdot K}{T}$$

We can also write:

$$\Delta G = - R \cdot T \cdot \ln V_g \cdot \rho L$$

Thus we may obtain:

$$\Delta S = \frac{\Delta W}{T} + R \cdot \ln V_g \cdot \rho L$$

Values of the heats of mixing obtained by gas chromatography may be found, among other places, in references 3, 7, 8, and 9; values for the entropy of mixing may be found in references 3 and 9.

Among others, Wolf and Beyer[10] have dealt with the determination of relative surface areas of adsorbents from gas chromatographic data. (See also Keulemans and Cremer, *Gas-Chromatographie*, Verlag Chemie, 1959.)

References

1. KOZUB, N., Diploma thesis for the Faculty of Mathematics and Natural Sciences, Karl-Marx University, Leipzig, 1959.
2. ANDERSON, J. R. A., *J. Amer. Chem. Soc.*, 1956, **78**, 5692.

References

3. ANDERSON, J. R. A. and NAPIER, K. H., *Aust. J. Chem.*, 1957, **10**, 250.
4. HOARE, M. R. and PURNELL, J. H., *Trans. Faraday Soc.*, 1956, **52**, 222.
5. KEULEMANS, A. I. M., *Gas Chromatography*, Reinhold Publ. Corp., New York, 1957.
6. KWANTES, A. and RIJNDERS, G. W. A., *Gas Chromatography*, ed. D. H. Desty, Butterworths, London, 1959, p. 125.
7. LITTLEWOOD, A. B., PHILLIPS, C. S. G. and PRICE, D. T., *J. Chem. Soc.*, 1955, 1480.
8. PORTER, P. E., DEAL, C. H., and STROSS, F. H., *J. Amer. Chem. Soc.*, 1956, **78**, 2999.
9. WHITE, D. and COWAN, C. T., *Gas Chromatography*, ed. D. H. Desty, Butterworths, London, 1959, p. 116.
10. WOLF, F. and BEYER, H., *Chem. Tech. Berlin*, 1959, **11**, 142.

Further Literature on Methods of Quantitative Gas Chromatography

HAUSDORFF, H. H., *Chemikerztg*, 1957, **81**, 392.
LEE, E. H. and OLIVER, G. D., *Analyt. Chem.*, 1959, **31**, 1925, describe the use of two substances for the internal calibration with substances not in the original mixture for wide boiling range mixtures.
LEROSEN, H. D., *Analyt. Chem.*, 1960, **32**, 444, describes the use of ethyl chloride as an internal standard for mixtures of light hydrocarbons.
SCHOMBERG, G., *Z. Anal. Chem.*, 1958, **164**, 147.
ZHUKHOVITSKII, A. A. and TURKEL'TAUB, N. M., *Zavodskaya Lab.*, 1958, **7**, 796, Analytical errors caused by incomplete separation.
ZHUKOVITSKII, A. A., TURKEL'TAUB, N. M., VAGIN, E. V. and SHVARTSMAN, V. P., *Dokl. Akad. Nauk SSSR*, 1954, **96**, No. 2, 303, Tailing in chromathermographic and thermal separations.
WEINSTEIN, A., *Analyt. Chem.*, 1960, **32**, 288, indicates some errors in gas solid chromatographic quantitative analysis that cannot be removed by calibration.
Also in *Gas Chromatography Abstracts* 1958, 1959, 1960 and later issues, ed. C. E. H. Knapman, Butterworths, London, under subject index No. 6.1.

Further Literature on Integrators

HALL, W. K., SILL, G. and WOLFE, C. L., *Science*, 1957, **126**, 821.
NOGARE, S. D., BENNETT, C. E. and HARDEN, J. C., *Gas Chromatography*, ed. Coates, Noebels and Fagerson, Academic Press Inc., New York, 1958, p. 117.
HALASZ, I. and SCHNEIDER, W., *Z. Anal. Chem.*, 1960, **175**, 94, Principle of a new type of non-electronic integrator with numerical registration.
HALASZ, I. and SCHNEIDER, W., *Gas Chromatography*, ed. R. P. W. Scott, Butterworths, London, 1960, p. 104.

Monographs on Gas Chromatography

BAYER, E., *Gas-Chromatographie*, Springer-Verlag, Berlin, 1959, 163 pp., 2nd edition, 1962, 324 pp.
ESAYAN, M. and ESAYAN, L., *Cromatografia Gazelor*. Editura Technica, Bucuresti, 1957, 158 pp.
KAISER, R., *Gas-Chromatographie*, Akademische Verlagsgesellschaft Geest und Portig KG, Leipzig, 1960, 223 pp.

KEULEMANS, A. I. M., *Gas Chromatography*, Reinhold Publ. Corp., New York, 2nd edition, 1959, 234 pp.
KEULEMANS, A. I. M., *Gas-Chromatographie*; translated and adapted by E. Cremer; Verlag Chemie Weinheim/Bergstr., 1959, 208 pp.
PHILLIPS, C., *Gas Chromatography*, Butterworths, London, 1956, 105 pp.
SCHAY, G., *Theoretische Grundlagen der Gas-Chromatographie* (Basic theory of gas chromatography), Deutscher Verlag der Wissenschaften, Berlin, 1960, 267 pp.

Gas Chromatography Abstracts 1958, 1959, 1960 etc. compiled and edited by C. E. H. Knapman, Butterworths, London (an outstanding collection of abstracts, issued annually).

Meetings and Symposia on Gas Chromatography

COATES, V. J., NOEBELS, H. J. and FAGERSON, I. S., *Gas Chromatography*, Academic Press Inc., 1958, 323 pp., Symposium of the Analysis Instrumentation Division of the Instrument Society of America, Michigan, 28–30 Aug. 1957.
DESTY, D. H., *Vapour Phase Chromatography*, Butterworths, London, 1957, 436 pp., 1st Symposium held by the Hydrocarbon Research Group of the Institute of Petroleum, London, 30 May–1 June 1956.
DESTY, D. H., *Gas Chromatography*, Butterworths, London, 1958, 383 pp., 2nd Symposium held by the Hydrocarbon Research Group of the Institute of Petroleum, Amsterdam, 19–23 May 1958.
FRESENIUS' *Zeitschrift für Analytische Chemie*, 1958, **164,** 1–218, Papers presented at the Symposium on the Analysis of Gaseous and Liquid Hydrocarbons by Physical Methods, Essen, 22–24 Jan. 1958.
Gas-Chromatographie 1958, ed. H. P. Angele, Akademie-Verlag, Berlin, 1959, 248 pp., 1st Symposium on Gas Chromatography, Leipzig, 9–10 Oct. 1958.
Gas-Chromatographie 1959, ed. R. Kaiser and H. G. Struppe, Akademie-Verlag, Berlin, 1959, 340 pp., 2nd Symposium on Gas Chromatography, Böhlen, 1959.
Gas Chromatography. Annals of the New York Academy of Sciences, 1959, **72,** published by the Academy, 559–785, 226 pp., New York Meeting, Autumn 1958.
Gas-Chromatographie, a selection of Soviet, Czech and Chinese papers, ed. R. Kaiser and H. G. Struppe, Akademie-Verlag, Berlin (in press).
SCOTT, R. P. W., *Gas Chromatography 1960*, Butterworths, London, 1960, 466 pp., 3rd Symposium held by the Gas Chromatography Discussion Group of the Hydrocarbon Research Group of the Institute of Petroleum, Edinburgh, 8–10 June 1960.

Sections of Books on Gas Chromatography

RÖCK, H., *Ausgewählte moderne Trennverfahren zur Reinigung organischer Stoffe* (Selected modern separation processes for the purification of organic substances), pp. 72–137, Dr. Dietrich Steinkopff, Darmstadt, 1957.
SOKOLOV, V. A., Methods of gas analysis in *Chromatographic methods of adsorption gas analysis*, Section 3, pp. 143–203: Gostoptekhizdat, Moscow, 1958.

INDEX